The
HOMING
INSTINCT

MEANING AND MYSTERY
IN ANIMAL MIGRATION

Bernd Heinrich

WILLIAM
COLLINS

William Collins
An imprint of HarperCollins*Publishers*
1 London Bridge Street
London SE1 9GF

WilliamCollinsBooks.com

This William Collins paperback edition published 2015

20 19 18 17 16 15
10 9 8 7 6 5 4 3 2 1

First published in Great Britain by William Collins in 2014
First published in the USA by Houghton Mifflin Harcourt in 2014

A catalogue record for this book is available from the British Library.

Book design by Lisa Diercks
Text set in Miller

ISBN 978-0-00-759585-3

Printed and bound in Great Britain by Clays Ltd, St Ives plc.

CONTENTS

PREFACE

ABOUT A DECADE AGO I STARTED PULLING TOGETHER BITS
and pieces on the "homing" topic and in 2011 had a book manuscript
scheduled for publication. I was then living at "camp" in Maine,
where I had done my fieldwork on bumblebees for years; lately my
work involved feeding ravens with cow carcasses in the winter, and
I then got interested in beetles that bury mouse carcasses in the
summer. Soon the topic of recycling of animal carcasses of all sorts
seemed more urgent than the scheduled book about getting to and
living in a particular place. So, I put writing about homing on hold.
By the time I again picked up my pencil, it seemed as though everything I thought of or had an interest in had, in one way or another,
started to have a bearing on home and homing. In the meantime
I had also been confronting personal issues of "homing," and they
seemed to take on increasingly similar forms to what I was reading
about in animals.

I had already left my academic position in Vermont and wanted
to return home to live in Maine, possibly in the home to which I
had bonded strongly as a child. I had planted a row of trees there
about thirty-five years ago. Those trees, now huge, brought back
many memories related to them. They reminded me of my father,
who had liked them, probably because he had strong feelings for a
row of chestnut trees that also lined the way to his old home in the
old country that he had often talked about. Because I had written

a book about my father, and not also one about my sisters or my mother, I had come into the disfavor of both. There had been parent-offspring conflict before my mother died, and then the house stood empty. So then there was also sibling rivalry over the estate. It seemed like being in a real-life situation of the sociobiology theories, in an almost perfect rendition of a naked mole rat colony where one of the family finds a big tuber, and the others claim it as theirs, and then a vocally assertive member establishes herself as the matriarch. I realized then that the difference between what can happen to a human and to a naked mole rat family is mainly one of terminology. This thus provided the topic of home and homing a much wider perspective.

INTRODUCTION

*Our passionate preoccupation with the sky, the stars, and
a God somewhere in outer space is a homing impulse.
We are drawn to where we came from.*

— Eric Hoffer

With all things and in all things, we are relatives.

— Native American (Sioux) proverb

I LEANED ON THE SHIP'S RAILING AT THE STERN, A TEN-YEAR-
old boy with virtually no notion of where my family might be going.
I heard the deep roar of the engines, the whine of the wind, and the
rush of the churning water. I felt adrift, as though carried along like
a leaf in a storm, feeling the rocking, the spray, and the endlessness
and power of the waves. I had no notion that we were among multitudes who had made hard decisions to court the great unknown, or
any clear idea of why my family had left the only home I'd known in
a forest in Germany. The only picture of what our new home might
be was that we might find magical hummingbirds, and fierce native
tribes armed with knives, bows and arrows, spears, and tomahawks.

Security for me was the memory of where we had come from,
specifically a little cabin in the woods and a cozy arbor of green
leaves that enclosed me like a cocoon where I could see out but no-

body could see in. It meant a feeling of kinship with the tiny brown wren with an upright stubby tail that sang so exuberantly near its snug feather-lined nest of green moss hidden under the upturned roots of a tree in a dark forest. I had in idle moments in my mind inhabited that nest. I found, too, the nest of an equally tiny long-tailed tit. This little bird's home was almost invisible to the eye because it was camouflaged with lichens that matched those on the thick fork of a tall alder tree where it was placed.

The ocean all around was a spooky void. But then, after several days at sea, a huge white bird with a black back appeared as if out of nowhere, and it followed us closely. I saw its dark expressionless eyes scanning us. It was an albatross. It skimmed close over the waves and sometimes lifted above them, circled back, and then picked up momentum to again skim alongside our boat. It followed us for hours, maybe even days.

The albatross was big and flew without beating its wings. Years later I wondered if, even in the featureless open ocean where so much looked the same every hour and every day, it may have known where it was all along. How do we find our home and recognize it when we find it? These questions were inchoate then, but given the examples of other animals, they put many ideas of home and homing in context.

Later, as a graduate student, I read that pigeons could return home to their loft even when released in unfamiliar territory, and that some other birds could navigate continental distances using the sun and the stars. There were few answers to how they did it. But I read about researchers at Cornell University who attached magnets to the heads of pigeons and got them all confused. Donald Griffin, my scientific hero (who had discovered how bats can snatch silent moths out of the air in a totally dark room that had wires strung all over the place), was releasing seagulls over forest where they could never have been before and then tracking the birds' flight paths by following them in an airplane. Most of his birds turned in circles before some of them flew straight, although why was not clear. Searching for a thesis problem to work on, I wrote to ask him if birds passing through clouds might keep in a straight line by listening to the calls flocks make while migrating. He replied in a long, thoughtful letter to let me know that this idea was too simplistic, and that one should not discount much more complicated mechanisms. That was excellent advice. I did not then have the means to solve any of these puzzles, but over the years I have kept in touch with the evolving field of animal navigation and its relevance to the need for a home.

For other animals and for us, home is a "nest" where we live, where our young are reared. It is also the surrounding territory that supports us. "Homing" is migrating to and identifying a suitable area for living and reproducing and making it fit our needs, and the orienting and ability to return to our own good place if we are displaced from it. Homing is highly specific for each species, yet similarly relevant to most animals. And the exceptions are illuminated by the rule.

The image of that albatross took on more meaning decades later, after I learned that the species mates for life and returns to the same pinpoint of its home, on some island shore where it was born, perhaps fifteen hundred kilometers distant. During the years when it grows to adulthood it may never be in sight of land. Seven to ten years after having left its home, it returns there to nest. It chooses to go there because of its bond. When a pair eventually have a chick in a nest of their own, each parent may travel over fifteen hundred kilometers of ocean to find a single big meal of squid, and after gathering up a full crop, it then flies home in a direct line; it knows where it is at all times.

. . .

The broad topic of homing subsumes many biological disciplines. In order to show the connections among all animals and us, I have interpreted the traditional use of an animal's "territory," or "home territory," simply as "home." We think of "home" primarily as a dwelling, but in order to be inclusive with other animals, I here consider their dwellings to be their homes as well. My application of the same terms to different species is deliberate for the sake of scientific rigor and objectivity, to acknowledge the continuity between our lives and those of the rest of life. I realize that this smacks to some of anthropomorphism, a pejorative term that has been used for the purpose of separating us from the rest of life. The behaviors involved in homing include drives, emotions, and to some extent also reason.

A home makes many animals' lives possible: home is life-giving and sought after with a passion to have and hold. We humans are not thinking much about "home" for animals when we confine them in cages devoid of almost everything they need except air, food, and water in a dispenser, or when we destroy the habitat that contains the essentials of home for many species. So I begin our exploration of home and its implications with the example of the common loons, *Gavia immer*, birds that may live for decades. The collaborative study by three biologists, Walter Piper, Jay Mager, and Charles Walcott, reveals how important home can be — enough for fights to the death.

Loons spend winters in the open ocean, but a pair migrate from it and across the land back to their home, a specific northern pond or lake, to nest along its shore in the spring and raise one or two chicks out on the water. Starting almost immediately after ice-out and almost until freeze-up, camp owners along a lake routinely see "their" pair of loons year after year. It had long been assumed that the same individuals return each year and live as monogamous pairs on their strongly defended home territory. Huge surprises were in store after 1992, when techniques (using a boat, a strong light, and a net) were developed to capture loons and mark them with colored leg bands to identify individuals. In a long-term study of a population of loons in Wisconsin in a cluster of about a hundred lakes, it turned out that *a pair* of loons indeed returned year after year to their home. However, they were not always the same birds. As expected, given their longevity and reproductive potential, there were many "floaters," those still without a home, and some of them routinely replaced members of a pair.

The floaters regularly visited different pairs at home at their respective lakes, and spirited vocal meetings resulted. These seldom led to fights, but they were not just friendly visits. These floaters were at first thought to invade others' home grounds in order to make "extra-pair parenting" attempts (which in males refers to extra-pair copulations and in females to egg dumping into the others' nests). However, DNA fingerprinting of the young loons from four dozen families produced not a single incident of extra-pair parenting. Instead, the visits by floaters were of an entirely different nature. They had an almost literally "deadly" purpose. The floaters were scouting — making assessments of both the worthiness of the others' real estate and the defensive capabilities of the resident males — to gauge the possibility for future takeovers.

Loons nest on the ground along shorelines, if islands are not available. Shorelines, if low, are risky nesting places, not only because of potential flooding in early spring, but also because they are within easy reach of raccoons, skunks, and other predators. Most birds test what is a "good home" by direct experience — success in raising a brood there. Or, like the loons, they assess the experience of others: whether or not chicks have been raised there. So, if the territory does yield young, a floater, who apparently finds this out through scouting, may risk a fight in an attempted takeover to remove the defending male (if he achieves takeover, he automatically gets to keep the female who remains on her home). But the floater can risk an attack only when he is about four or five years old. When he is in prime physical condition and has much to gain — namely, a potential lifetime of home ownership — it makes sense to risk the battle. The older territory owner might then fight to the death, presumably because he has much to lose, and almost no chance to gain another home.

Loons may seem extreme in the lengths to which they go to secure a home, as do other birds that risk the hazards of migrating thousands of kilometers. Yet, in the movie I watched with rapt attention on board ship on the way to America at age ten, people on ponies shot arrows at others on a wagon train. All were emotionally charged, because each was fighting for something sacred, and therefore each was willing to risk his life, for defending or wanting a home.

We have learned much about thousands of animal species that twice annually risk their lives to migrate to an exact pinpoint, such as an oceanic island in the case of the albatross or a pond in the vastness of a continent in the loon's. They open a window with a broad view onto our unending quest into the mysterious minds of animals, and in the process they illuminate our own. How does one tie the vastness that includes other animals, and so much that affects us personally and socially, together into a story?

Writing this book reminded me of when, after riding more or less unconscious in the slipstream of history for over fifty years, I started putting down stones chosen from a vast array of differently shaped fieldstones to build a house foundation. I couldn't chop them to size or knock the edges off so that they would fit into the inevitable empty spaces to make neat connections. Nor did I want to shape them, like bricks, to make a tidy but artificial structure. The stones found in nature, like facts, are endlessly numerous, wild, and complex. As the famous British geneticist and evolutionary biologist J.B.S. Haldane quipped, "My own suspicion is that the universe is not only queerer than we suppose, but queerer than we can suppose." I hope to give here a view of some of the "stones" of homing, and their origins, and how they apply in real life.

Homing is central to many aspects of our and other animals' lives. To understand the meaning of home, like any other phenomenon, it helps to step back and see from another's world. Animals give more than just clues to the why and the how of homing. They show what is possible, what has been tested, and what has worked over millions of years of evolution.

In this book I cannot hope to provide an in-depth treatment of any issue. My viewpoint is wide, and admittedly personal. Thousands of scientific references are possible for any one topic, and I am not an expert on any topic. Those references I cite are in no way meant to be the specific last word on the topic. They tend to be those I am most familiar with. I apologize for all of the amazing stories and all the details that I have left out. I have tried to speculate freely, and I hope that this will open discussion, not close it.

PART I

Homing

A skull I have lying on my desk is as big around as a somewhat flattened coffee mug, but it comes to a point at the front and has two large eye sockets. The bone on the top of the head is sculpted in ridges and furrows and looks like weathered stone. Seen from the back, the skull has a backward-opening hole on each side. The holes anchored the animal's powerful jaw muscles. Between them is a much smaller hole. I am looking through this hole (foramen magnum) for the brain cavity, but this large head contains scarcely any brain at all, just an extension of the dorsal nerve cord that runs up from the neck.

It's the skull of a snapping turtle, a female who met her

end as she was crossing the road to dig a hole in the gravel to lay her clutch of eggs. As in previous years, she had come from her bog a mere hundred meters away.

The local snapping turtle is no great wanderer. But two sea turtles — the green turtle, Chelonia mydas, *and the leatherback,* Dermochelys coriacea — *are renowned for migrations spanning entire oceans. The leatherback as an adult feeds in northern cold waters but nests on tropical beaches. It can weigh up to a ton, and although the skull of one that I had the opportunity to examine was as big as a basketball, the cavity holding its brain was, like the snapping turtle's, only a slight expansion at the end of the vertebral column. It could barely have held a walnut. The green's could have held two hazelnuts. Their skulls hardly differ from that of any turtle, whether of a species alive now or one that lived 215 million years ago, at a time at least three times more distant than that of the last dinosaurs.*

The minute dimensions of some animals' brains are as astounding as the homing capacity of some of their owners. Like albatrosses, sea turtles of various species lay their eggs in colonies with others of their kind on specific ocean beaches. *After the young hatch from the eggs buried in the sand, they head for the water and spend years at sea. They may travel thousands of kilometers, and then, a decade or two later when they are ready to lay their eggs, they return to their birthplace. They mate in the water nearby, and the females then come ashore to dig their nest holes in the sand and to drop in their eggs. How are they able to find their old home after years of wandering in the vastness of the oceans, when we, if taken blindfolded to and then released in unfamiliar woods, would, despite our highly sophisticated massive brains, be as likely to head off in a wrong as a right direction? To get around in unknown territory most of us need a map with which to find at least one known fixed feature that we can both see on the ground and locate on that map, and a compass.*

What knowledge and what kind of urges does it take for some birds to fly nonstop for nearly ten thousand kilometers, spending all day and night on the wing, until their body weight halves as they not only burn up all of their body's food stores but even sacrifice muscle, digestive tract, and other entrails — almost everything except their brains?

CRANES COMING HOME

If feeling fails you, vain will be your course.
— Johann Wolfgang von Goethe, *Faust*

MILLIE AND ROY ARE A PAIR OF SANDHILL CRANES THAT STAY for most of the year in Texas or Mexico but travel north in April and have for at least fifteen years nested and raised their one or two offspring, known as colts, in a small bog in the Goldstream Valley near Fairbanks, Alaska. Their home is adjacent to the home of my friends George Happ and his wife, Christy Yuncker. George was an insect physiologist and chairman of the Zoology Department at the University of Vermont where I was hired in 1980, and he later moved to the University of Alaska and the land of the Iditarod, where the two built their home in the wild land near Fairbanks. They invited me to visit them and "their" cranes, and I was eager to do it.

The thousands of square kilometers of central Alaska's permafrost-covered taiga consist of stunted blue-green spruce and white birch, with a groundcover of green-yellow moss and twiggy Labrador tea whose evergreen leaves curl at the edges and have a soft

Portrait of Millie and Roy

beige fuzz on their undersides. Chalky lichen and small shiny cran-
berry leaves decorate a thin black soil overlying the permafrost that
can extend thirty meters down. In this expanse, there are many
bogs or pingos, which are the result of an ice dome (groundwater
that freezes into an upwardly bulging ice lens) that has melted and
created a depression where a pond or a lake is then formed. After a
few centuries, a floating mat of vegetation grows in from the edges
to create a floating bog. Such pingo bogs have become the favorite
home sites of sandhill cranes.

George and Christy's pingo in the Goldstream Valley is, like
others, clear of trees but surrounded by stunted black spruces. It
is the home site not only for the crane pair but also for Bonapar-
te's and mew gulls, pintail and mallard ducks, and sometimes
horned grebes. I intended to arrive several days in advance of
the cranes' anticipated return, to try to watch their homecom-

ing. Surely this return during the first week in May would be a big event in their lives, and I wanted to see their reactions to their old home.

Millie and Roy had last been seen as they left the bog in the previous year, on September 11, 2008, for their southward migration. They had been delayed from their normal end-of-August departure date because Oblio, their colt, had a leg injury that prevented him from being ready in time for the family flight to western Texas. Waiting for him saved his life; we know that young cranes, as well as geese and swans, learn the route between wintering and breeding homes from their parents. The proof and the implications of the necessity of the young to be able to follow their parents, or alloparents, in order to migrate were perhaps most convincingly demonstrated by William Lishman after he first played parent to hand-raised geese that he later led as a flock with an ultralight aircraft. He also led sandhills in this way. Finally he led a flock of whooping cranes from their breeding grounds in Wisconsin to establish new homes for them in Florida. However, nobody as far as I know has been able to follow wild birds, and my chances of seeing Millie and Roy touch down for the first time on their arrival this year might be slim. But I felt it was worth a try.

Cranes, like other large birds, grow slowly. It takes them thirty to thirty-two days to incubate their two eggs, and another fifty-five days for their (usually one) colt to be able to fly well enough to migrate. This far north there is only a narrow window of time for cranes to breed successfully, especially for those that fly even farther to breed, as some of those wintering in Mexico do, in Siberia. If late in arriving, they waste their effort of migrating the thousands of kilometers north. If they are too early, snow and ice cover all food sources. This year had been a winter of heavy snows in central Alaska. Even the boreal owls were starving from their inability to reach the voles under the snow. By late April, when I arrived, the woods around Millie and Roy's home bog were still under at least half a meter of snow, and the cranes had not yet shown up.

I would have liked to fly with the cranes on their homeward journey, but the best I could do, apart from trying to beat them to their destination, was to see a piece of their flight path. My transcontinental flight of 3,872 kilometers from JFK Airport in New York was followed by a direct flight on April 23 from Seattle to Fairbanks, and I spent most of the three and a half hours of the 2,467-kilometer flight from Seattle north in the Boeing 737–800 with my face pressed to the window, trying to see like a crane. How did the cranes navigate and negotiate their five-thousand-kilometer journey from Texas or Mexico to come home to their own pingo out of thousands of others scattered throughout the vast and seemingly unending Alaskan taiga?

The cranes arrive lean at their main staging area, at the Platte River in Nebraska, and stay three weeks to gather reserves for their continuing journey north. When ready, they gather with thousands of others and wheel high in the sky into giant "chimneys," to travel together on their common journey. Once in Alaska, they take separate paths to their individual homes, and a third of them fly beyond, to their homes in Siberia.

We had scarcely lifted off in Seattle when we passed over white-capped mountains with knife-edged ridges, dark forested valleys, and peninsulas surrounded by blue-gray water. An hour later, cruising at about eight hundred kilometers per hour at eleven thousand meters, there was ever more of the same — white mountains as far as the eye could see. Another hour — it was still the same. To me, barely a feature stood out from the jumble of endless peaks that melded into each other, and the vast mountain scape was broken only by frozen lakes glinting in the evening light. And so it continued for yet another hour. When we started our descent to Fairbanks,

I saw oxbows of meandering rivers, and finally the thin thread of a road.

Cranes, swans, and geese travel south in the fall as family groups. On their way, the young learn the route they will later take north in the spring, to come back to try to settle near where they were born. What they see and remember seems astounding. I might, with intense concentration, memorize a tiny portion of the way, perhaps around this or over that mountain. But these cranes come not from my point of departure, the state of Washington, but from considerably farther south. (Four cranes from the Coldstream Valley that the Alaska Department of Fish and Game had equipped with radio transmitters ended up in various parts of Texas in the winter.) I could never retrace even my own much shorter flight route from Seattle, even if I were to return the day after having flown over it, much less a half-year later. What are the cognitive mechanisms that allow the birds to do this?

Day after day for almost eight months now there had been no crane at the pingo. For most of that time the ground had been under a deep layer of snow that locked any food out of reach. What would happen if, after their long-distance flying, the pair were to arrive at their home and find the bog still under snow and ice with no cranberries to be found and no voles to catch? How much can cranes afford to gamble in order to try to come on time, or even early?

It was only in the last week of April, after another snowstorm, that the weather suddenly warmed, and just then, on the 24th of April, on my first morning, we heard a crane in the distance. Still, no cranes landed on the pingo on the 25th, 26th, 27th. But the next morning at dawn I awoke to the loud and penetrating trumpeting calls of a single crane. These metallic sounds are unearthly; as Aldo Leopold wrote in his "Marshland Elegy," they evoke "wildness incarnate." On and on this bird shattered the dawn's stillness, and I ran out to look. But the bird was then distant, and the sound kept shifting position, so I presumed it was flying around in great circles, possibly looking for a patch of cleared ground; the mossy floor of the nearby stunted spruce forest was still covered in deep snow.

That evening we sat down to supper by the window facing the wide-open panorama of the pingo in front of us. The sun was still high. We were just polishing off the last of our freshly grilled salmon when we looked up to see a crane with spread wings gliding down for a landing. Its long thin legs touched down gently on the still-thick ice of the pond. It bugled and sprang up half a meter or so, unfurling its long neck with beak held skyward and with extended wings at the apogee of its graceful leap as if to catch itself in the air to prolong this moment. It looked like a physical embodiment of joy and excitement. The crane kept leaping, all the while continuing its stirring bugling. Cranes don't do this every time they land; this was indeed a special landing. Finishing its dance, the crane started to walk in a contrastingly slow and deliberate manner, thrusting its head forward and up with each step, and at the same time opening its bill and making a very different, trilling, call.

The crane walked and leaped in several more repetitions of its dance before eventually lifting off with even wing beats to sail off in the same direction from which it had come, its haunting cries growing faint in the distance. Two hours later the (same?) crane came again, but this time it circled the bog only once before leaving, continuing its calling. It had given the impression that it was glad to be back but was at the same time agitated and looking for something; a mate? In previous years Millie and Roy had always returned as a couple, George and Christy told me, so lacking positive identification, we were skeptical that this was either of them.

I had barely gone to bed that night when I heard another one or possibly two cranes calling excitedly, while a third seemed to answer from the distance. I jumped up and rushed outside to look:

Crane pair coming home

a *pair* were walking from one end of the bog to the other. George and Christy were up also, and for the next hour, until it got dark at 11:15 p.m., we watched. One of the cranes stood tall and extended his head and neck forward and up, reminding me of the dominance display of a male raven. This was Roy. The second one, whom I would soon identify by her walk and talk and narrower white face patch, held her neck and head in a more downward curve like a heron's, projecting her head slightly forward with each step. She started to pick up cattail fronds and grasses and then to deposit them, sitting down briefly on the materials. Was she encouraging her mate by suggesting to him that it was time to start nesting at one or another of these potential nest sites? I could hardly wait to see what would happen next.

Another clamor awoke me in the early dawn, around 4:00 a.m.

Cranes dancing

Cranes strutting

There had been a heavy overnight frost and the two cranes were standing on the ice in the middle of the bog. Both Millie and Roy tossed their heads up in a quick motion, their bills opening wide during each call. We heard what sounded like a hammer hitting a metal bell or drum, as he opened his bill once to make each call and she chimed in at the same time but called and opened her bill twice to make two short similar cries of a higher pitch. It was a composite call made by both together; a duet. A third, distant crane responded. The distant calls, and the pair's duet, were repeated back and forth, often and loudly.

The vigor of the pair's unison calling was still palpable, even when the first morning light lit the sky and silhouetted the black spruces, and when I thought about the enormous effort they had invested to get here, I realized what was at stake: home ownership. The pair's loud clanging calls attracted no others flying in from the distance; instead, the calls are a vocal "no trespassing" sign, one leaving no doubt that any potential challenger would be facing not just one bird, but a united, cooperating *pair*.

I watched the pair for another hour. She by then occasionally fluffed herself out and, as she had done the day before, continued periodically to squat where she had pulled at or dropped sedge and other potential nesting material.

The pair continued their slow, deliberate steps that morning while meandering from one end of the bog to the other, as though inspecting every square centimeter of it, and at 9:00 p.m. we saw Roy jump high and with outstretched wings dance by himself out on the ice. Millie dashed by him with fluttering wings, to round out a mutual performance. A little later we heard a purring call as she spread her wings to the sides and stood still. Roy, with outstretched neck and elevated bill, jumped onto her back and, balancing himself with a few wing beats, mated. He dismounted after a couple of seconds. Both then bowed to each other and continued their walk.

They were home, intended to nest, and had now sealed the deal.

Still, the lone bird that had tried to intrude on their turf did not easily abandon its intended claim. But why would this crane want a claim without a mate to nest with? Was it Oblio, their grown colt from last summer? Was he, now that Millie and Roy were re-nesting, finally being "thrown out of the nest" by his parents, who no longer tolerated company of any sort? I suspect this was the case. The offspring of most birds have a strong attachment to home. This emotion is a biologically relevant drive, because home is where reproduction has proven to be successful. But the young have a life-time ahead of them, and if for some reason the parents don't make it back home, the offspring could inherit their territory. If the par-ents do come back to reclaim their home, at least nearby territory would be more like the old home than the far-off unknown. Even if a bird is without a mate, finding a suitable territory is often a pre-requisite to getting one.

Shortly after we had breakfast the next morning, the lone crane flew over once again, and the pair immediately launched their syn-chronized duet as a vocal challenge. This time, although the lone crane did not land, the pair jumped into the air and chased it until all three were out of our sight and hearing. But the pair returned soon and then again performed several nest-building probes. They again mated in what would become a routine for the next several days: at least once in the morning and once in late afternoon or at night.

When I first saw the cranes the next dawn, each was standing on one leg with its head tucked into its back feathers. Ice had again solidly covered their pond, after having partially melted along the edges the day before. Both Millie and Roy seemed to be asleep, al-though he, balancing himself on one leg, occasionally reached up with the other to scratch his chin and head with the toes of the foot. But whenever a crane called from the distance, both their heads shot up instantly, and they renewed their spirited in-unison call, she making the two short notes and her mate making one, at a lower pitch.

We had so far not seen them feed. Indeed, it was hard to imagine that there was food available in any case. If lucky, they might by now catch a vole or find a few of last year's cranberries, but this year that would probably not be likely until days later when the snow would melt. A crane's large body size requires much more food than a small bird's, but that same large size is an advantage in tiding them over during lean periods, and thus to return to their homes before the anticipated flush of food becomes available.

As it got lighter on this dawn, the cranes soon had company. A pair of swans and then a small flock of five Canada geese flew by. A robin sang. By about 7:00 a.m., the crane pair became animated as well, walking over the bog while picking here and there, and again mating.

We did not see the pair most of that afternoon. The lone crane, seemingly to take advantage of their absence, again flew in and this time landed, looked around, and repeatedly made a trilling call. But in about fifteen seconds the pair flew in as if out of nowhere, and one of them sped over the ice as if to attack the interloper, which immediately flew off. The pair gave chase, and all three disappeared from sight. In several minutes the single bird returned. Again the pair came and caught up to the lone crane before it had a chance to take off, and this time they attacked it viciously, in a flurry of flailing wings.

The pair had by now, after the third day, won the major part of the battle. Barring accidents, they would within days lay their two eggs and go on to raise their colt. Normally both eggs hatch, but as in some eagles and vultures, usually only one chick survives, probably because one gets fed less and then weakens and eventually starves. Presumably through evolutionary history, for the fast growth required to reach full development and readiness to migrate by August, there has not been enough food to raise two colts at once. One might suppose the cranes could simply lay only one egg, but sometimes an egg does not hatch, and the second is insurance.

The pair seemed more animated after their last fight with the lone intruder, and by evening they again mated (for at least the third time that day). In most birds one mating is enough to fertilize the eggs. Perhaps several matings are insurance, but this seemed more than enough for insurance. Perhaps, like their dances, mating is additionally part of their bonding ritual.

A pair of mallards, and then a pair of pintails, arrived in the evening, and the ducks swam next to each other near the cranes at the edge of the pond, where some open water had reappeared during the day. The cranes ignored them and again walked in their stately manner back and forth across the ice of the pond, and now they pecked in the low vegetation being exposed along the edges. They were by now finding overwintered cranberries exposed by the melting snow.

On the evening before I would leave for my journey home, Christy and George hosted a potluck party. Shadows fell on the white frozen middle of the pingo as the western sky turned yellow and orange and the spruces became dark silhouettes. A pair of pintails again landed in the open water along the pond's edge. The cranes were standing, each on one leg, their heads tucked into their back feathers. People crowded around the spotting scope in the living room, watching them occasionally shift position, lower a leg, poke a head out to look around. Suddenly the person then at the scope erupted with an exclamation: "They are mating!" She had seen the male approach the female with her spread wings, mount, flutter, and jump off. The pair had bowed to each other. Suddenly many people crowded around the scope to watch.

Why, I wondered, would anyone, or almost everyone, want to

watch cranes mate? Why was nobody interested in watching the mating activity of the two ducks, or of the numerous redpolls? Could it be, I wondered, because we feel a closer kinship with cranes than with other birds?

Cranes are similar to us in many ways. Some are nearly as tall as a person. They walk on two long legs like us, albeit with a much more graceful and deliberate gait, so that they remind one of a caricature of a gentleman or an elegant woman on a leisurely stroll. The sandhill crane's red bald pate and sharp yellow eye add to the caricature. Cranes form lifelong pairs and stay together as families, but they are also gregarious and join up into large groups. They form a strong attachment to their home. They not only make music with trumpeting calls that sound like bugles, but they also dance, and do so on various occasions.

All of the fourteen species of the world's cranes dance. Crane dancing involves running, leaping into the air, flapping the wings, turning in circles, stiff-legged walking, bowing, stopping and starting, pirouetting, and even throwing sticks. Dancing is primarily done by pairs and presumably functions in cementing pair bonds and/or synchronizing reproduction. But it can also be induced at any time, and it stimulates other cranes to dance. Even the young colts perform some of the species' dance. Possibly it serves as practice and could be motivated by the same basic emotions of joy that are an indicator of health important to mating.

Cranes' dances often stimulate humans to dance as well and have been mimicked in many cultures all over the world where cranes live. Crane dances were performed by ancient Chinese, Japanese, southern African, and Siberian people. If not emulated, cranes are admired. In the Blackfoot tribe of Native Americans of northern Montana, the last name "Running Crane" is common.

Nerissa Russell, an anthropologist, and Kevin McGowan, an ornithologist from Cornell University, revealed that eighty-five hundred years ago at a Neolithic site in what is now Turkey, people probably performed crane dances using crane wings as props that were laced to the arms. Furthermore, someone of these people apparently hid a single crane wing in a narrow space in the wall of a mud-brick house along with other special objects (a cattle horn, goat horns, a dog head, and a stone mace head). Russell and McGowan also found evidence that vultures may have been hunted for their feathers for presumably a much different costume worn as well for a ceremonial purpose. The authors inferred that the cranes were linked with happiness, vitality, fertility, and renewal (since they arrived in the spring). While the crane dance was one of life and birth, and possibly marriage and rebirth, the vulture dance was associated with death and perhaps return to the afterlife.

Russell and McGowan believe that the crane wing interred in the wall of the house was never intended to be seen. It was a symbolic object related to marriage and construction of a new home and may have been coincident with a particular human marriage and home-making. The associations among dancing, pairing, and raising young and home would have been natural for people who saw cranes return to their home ground, just as I had seen Millie and Roy do. Seeing the close parallels in the biology of the birds with their own lives, and understanding the cranes' dancing as helping to make or cause the good things that followed, Neolithic people would have been compelled to symbolically emulate the crane dance of homecoming and of new life.

BEELINING

Observation sets the problem; experiment solves it,
always presuming that it can be solved.

—Jean-Henri Fabre

CRANES FLY AN ENORMOUS DISTANCE TWICE ANNUALLY, BUT relative to their size, bees also fly huge distances — up to ten kilometers — and the foragers may perform such trips hourly. We can experiment with them to find out how they navigate. What we know about bee homing so far is nothing less than astounding, and it is built on a long history of research, primarily pioneered by the imaginative experiments dreamed up and performed by an Austrian named Karl von Frisch and his colleagues that date back over a half-century. Arguably, our knowledge dates back still further to early American frontiersmen trying to find bees' treasure troves of honey.

In 1782, Hector St. John Crèvecoeur, a writer and farmer from Orange County in New York State, wrote:

> After I have done sowing, by way of recreation, I prepare for a
> week's jaunt in the woods, not to hunt either the deer or the bear,

as my neighbors do, but to catch the more harmless bees. . . . I proceed to such woods as one at a distance from any settlements. I carefully examine whether they abound in large trees, if so, I make a small fire on some flat stones, in a convenient place; on the fire I put some wax; close by this fire, on another stove, I drop honey in distinct drops, which I surround with small quantities of vermillion, laid on the stones; and I retire carefully to watch whether any bees appear. If there are any in the neighborhood, I rest assured that the smell of burnt wax will unavoidably attract them; they will find the honey, for they are fond of preying on that which is not their own; and in their approach they will necessarily tinge themselves with some particles of vermillion, which will adhere long to their bodies. I next fix my compass, to find out their course — and, by the assistance of my watch, I observe how long those are returning which are marked with vermillion. Thus possessed of the course, and, in some measure the distance, which I can easily guess at, I follow the first, and seldom fail of coming to the tree where those republics are lodged. I then mark it [presumably with his name to claim ownership].

James Fenimore Cooper, author of the Leatherstocking Tales of the American frontier, of which *The Last of the Mohicans* is probably best known, in 1848 published the novel *The Oak-Openings; or, The Bee-Hunter*. Here Cooper depicts a different, perhaps more reliable method than Crèvecoeur's of the frontier activity that came to be called "beelining." Cooper's story takes place during July 1812, in the "unpeopled forest of Michigan," where, due to the Native Americans' lighting periodic fires to clear the ground, there were many flowers among the scattered oaks. This was ideal honeybee habitat, and here the bee hunter Benjamin Boden, nicknamed "Ben Buzz," practices his art. Ben captures a bee from a flower by placing a glass tumbler over it and sliding his hand underneath. He then places the tumbler with the captured bee on a stump next to a piece of filled honeycomb. He puts his hat over the tumbler and the honeycomb so the bee will not be able to escape. He waits as the bee, stumbling around in the dark, eventually finds the honey. Once it is preoccupied with imbibing the honey, it quits buzzing, and the silence is the signal for Ben to remove the hat and then the glass, as the bee will stay to finish its feast and will fly up, circle the honeycomb, and depart directly toward its nest. He then follows the bee to the tree, chops it down, and is rewarded with just over one hundred kilograms of honey. Easier said than done.

American honey hunters eventually added refinements to their beelining techniques. The main improvement was the invention and use of a "bee box," a small wooden box designed to catch a bee and get it "drunk" on a hunk of honeycomb. It was used in Maine when I was a kid (I still own mine). George Harold Edgell, a lifelong bee tree hunter from New Hampshire, wrote in 1949 in a pamphlet titled *The Bee Hunter* that "one's first task is to catch a bee and paint its tail blue" and "this must be done gently [because] bees do not like to be painted. To paint a bee, it is best to wait until it is eagerly sucking up a thick sugar syrup and is too pre-occupied to notice."

By 1901 Maurice Maeterlinck, the Belgian playwright and Nobel laureate in literature, described in *The Life of the Bee* his scientific experiments on bees that were individually identified with daubs of paint, from which he deduced that these insects could communicate their discoveries of food bonanzas to hive mates that would then navigate directly to the food. However, American woodsmen not only had used similar methods, but had also, through their bee-lining, already gleaned that same surprising insight into what the bees could do. Maeterlinck credited his American predecessors for their discoveries and wrote, "The possession of this faculty [to communicate food locations to hive mates that then can navigate to the food] is so well known to American bee hunters that they trade upon it when engaged in searching for nests."

Although early American woodsmen, whose lives depended almost directly on the knowledge gained by close contact with nature, were beelining devotees who had deduced that honeybees recruit hive mates, it would remain for Karl von Frisch to unravel the marvelous story of *how* the bees communicate within the hive. He earned the Nobel Prize in Physiology or Medicine for this work. I feel lucky that a Maine neighbor, Floyd Adams, took me beelining when I was eleven years old, and that when I was a teenager, my father gave me an inspiring little book by von Frisch entitled *Bees: Their Vision, Chemical Senses, and Language.* It explained the experiments that he and colleagues had performed. They were mesmerizing because they connected the practical experience of beelining in the Maine woods with the imaginative power of a scientist who had penetrated into the core of the bees' world, their hive, their home.

Floyd's family's home was the farm four hundred meters down our dirt road. It was populated by chickens, geese, cows, pigs, plus all the other usual and unusual wildlife that lives in a place with a tolerance for disorder. Along with Floyd, my companions were the four Adams boys, Butchy, Billy, Jimmy, and Robert, an in-law of theirs. Floyd, a dark-haired, mustachioed, wounded Marine Corps veteran recently returned from the Pacific, had a bad limp and a thirst for Black Label beer. Leona, his blond, petite wife, appreciated his fondness for honey but less so his taste for beer. He and the "boys," after a hot day haying, sometimes went fishing on our nearby Pease Pond in the evening, but in August our big draw was always the beelining.

After we found a bee tree, we carved our initials into the bark to proclaim ownership (property lines were irrelevant with regard to bee trees; finders keepers was the rule), and at some convenient time we returned with crosscut saw, axes, wedges, a beehive, and pails and kettles for honey. Getting part of our living from the land was fun, and it meant understanding and using the bees' homing behavior to find their hollow trees in the forest and resettling them into a new home, which we brought back to the farm and set up at a window in the attic of the house.

Fast-forward to a quarter-century later: My nephew Charlie Sewall and I are in a patch of goldenrod blooming in a pasture where each fall the wild honeybees gather nectar to top off their honey stores for the coming winter. We start by capturing a single bee in our bee box, a simple four-sided wooden box that has a ten-by-fifteen-centimeter piece of honeycomb with sugar syrup filling out the bottom. We dab the box with a drop of anise for scent and capture our bee by holding the box under her after she has landed on a flower and then slapping the box cover over her. At first the captive buzzes in the box trying to escape, but the buzzing stops when she stumbles onto the sugar syrup and starts to tank up, which will take her a minute or two. We then remove the cover and set the open box onto a pole that reaches to just above the tips of the goldenrod. We gently daub her with a spot of paint while she is absorbed in sucking up syrup, as I remembered Floyd doing. We then hunker down into the goldenrod and wait as she continues sucking up her newfound sweets that she will soon share with her hive mates. After about two minutes, her honey stomach is filled. She crawls out onto the edge of the box, stops to wipe her antennae with her front feet, lifts off, and flies back and forth downwind of the box. We duck lower to keep her silhouetted in sight against the sky as she starts flying loops, which become increasingly wider and oriented in one direction. Finally she straightens her flight path and takes off, making a "beeline" into the distance. Knowing that nobody in that direction keeps bees, it's clear that she is on her way to a bee tree. She will soon be back with others, and we then consult our wristwatches to time her trip. A bee flies about four hundred meters a minute, and it may take her three to six minutes in the hive to regurgitate and unload her honey stomach's contents into the mouths of begging, receiving bees.

We settle down and wait, and after perhaps ten minutes or less a bee suddenly appears and makes very rapid zigzagging flights just downwind of the box. The sound of her fight has a higher pitch than that of the bees foraging on the nearby goldenrod flowers. This means that she is more motivated and has a higher body temperature because of the rich food she is expecting. She settles into the box and starts imbibing the syrup. More bees will come soon, and when they get near our bee box, they will be guided in by the scent of the anise that marks the spot. After they tank up, we watch their flight directions.

If the food is in the immediate home vicinity, the bee does a "round dance" on the honeycomb when she returns to her home. She repeatedly runs in small circles while shaking her abdomen, and she regurgitates small samples of her find at intervals during her dance. If they become motivated after receiving information about the quality and scent of the food advertised, her hive mates leave the hive and search for the advertised food. If it is beyond a few hundred meters, the bee alters her dance to also contain information concerning location. The distance of the journey to the food is proportional to the duration of the waggle runs, and the angle of the straight runs with respect to the vertical direction informs the bees in what direction to fly when they leave the hive. If the straight run is in the *up*-direction on the honeycombs (which always hang vertically in the hive), the food source is in the direction toward the sun. If the food location is, for example, at an angle of ten degrees to the right of the vertical, the food direction is ten degrees to the right of the horizontal component of the sun direction when the bee would fly from the hive. Thus, her behavior is a symbolic representation in body movements of the flight to the food.

The first steps in the evolution of recruitment likely involved simple alerting signals in or at the nest entrance before takeoff. Other bees could have followed those signaling bees, probably by scent, for at least a short distance in flight. Through a few million years, the alerting likely became modified to take on an ever-greater leading function by bees flying in an ever more conspicuous manner in the direction of the food, so that followers could start off flying with ever greater accuracy in the right direction. These flights, later in the evolutionary progression, were eventually restricted to a buzz run directly on the top of the combs, but still in the food direction. We can infer this, because such "primitive" recruitment is still found in some tropical honeybee species that have their combs in the open air, where this mechanism makes sense. But open-air homes, though convenient for such communication of food location, were vulnerable to predators and also precluded the bees from living in huge areas of the globe, those with cold climates.

Homes in hollow trees allowed the bees to live in areas where they would otherwise be excluded because of cold and/or nest predators. But in such safer homes the combs hung from the *roof* of a cavity and left no horizontal dancing platform, and additionally the "dance floor" was now in darkness, so bees could not point *directly* toward the food. Even if they could, they would not be seen. But a breakthrough for indicating horizontal directions on vertical surfaces became possible after some bees started using the *hanging* flat surfaces of combs as their dancing platform while indicating the sun's location as the up or "toward" direction in their dance. Additionally, tactile rather than visual orientation became predominant for recruits in reading the code within the nest.

. . .

It is amazing enough for an animal to be able to navigate to a location it has never been to before. But some ants do something even more amazing. In North Africa, desert ants live in underground homes where they are protected from the heat. But they must venture out onto the searing surface periodically to forage by scavenging on heat-killed prey. The ants are fast runners that have evolved a very high tolerance for heat. Still, at times it is a matter of life and death even for them to make it back to their cool underground home; they cannot afford to wander on the sand surface for an extended time without access to their shelter to cool down and replenish body fluids. This is where their homing ability comes in; they may have zigzagged in all directions to find a heat-killed insect, but after finding one they must make a straight "ant line" directly back home. This begs the question, Since they are often on a featureless plain and have not kept a steady course, how do they know in what direction to head home?

If one captures bees in one pasture and releases them in another, they usually depart in the direction they would have flown from the original field. That is, they act as one would expect if they do not realize that they have been moved to a new location. Rüdiger Wehner and his colleagues at the University of Zurich came to the same conclusion about desert ant homing in their lifelong experimental studies. The ants use the sun as a compass, but a compass is not enough; the ants, when released from a point they had not themselves traveled to, like the bees caught in one pasture and released in another, apparently got lost.

For homing you must know where you are on "the map" before you head off in the correct direction. The desert ants can return home, but only if they *walk* to where they find themselves. Wehner concluded that the ants' homing mechanism involves somehow calculating where they are at all times, probably in measuring distance by keeping a kind of count of their steps, and also keeping track of the angles of their direction from their home relative to the sun's location. These were not mere speculations, but a hypothesis tested in painstaking experiments that entailed altering the ants' perception of the sun (holding filters over them that varied the direction of polarized light that they, like bees, use in orientation) and altering their stride length (altering their leg length by gluing on extensions) to find out what information they valued and how they used it. Presumably bees could also have a similar "map sense," and Randolf Menzel, a neurobiologist in Berlin, was trying to find out how it might work.

Menzel runs the large and active Institute of Neurobiology at the Free University of Berlin, and one of his projects was the burning question of how honeybees seem to find out where they are in order to be able to go where they want to be. Honeybees are suitable animals with which to study this problem because, like ants, you can count on their motivation to return home after they are loaded with food.

We can't look into a bee's brain and determine what it knows and what it wants. However, clever experiments based on the bee's natural history permit inferences. We can determine, for instance, where a bee perceives herself to be relative to her hive. If a bee regularly visits a feeding place, she knows where she is, because she always flies off in a straight line from it back to her hive. If we then remove either the hive or the feeding spot, she circles in the area where her target had been. We know what she is looking for, because when we provide the hive and/or the feeding station within the area where she circles, she quickly finds it. But suppose we capture our bee at the usual feeding station after she tanks up on honey or syrup, put her into a dark box, and then carry her "blind" to a place she has never been. As mentioned, most bees will then make a beeline in the same direction they had normally flown to return

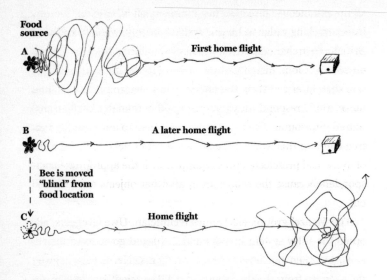

Bee flight paths. A. A bee's first trip from a flower patch or bee box back to its bee tree (hive) begins with an orientation flight. B. Later trips are more direct. C. After a bee has been transferred while "blindfolded" to a new spot, she acts as though she perceives herself to be still at the same place as before.

to the hive. They will fly as far as before but find no hive there. Yet, they usually eventually *do* make it back home. How do they find their way? What do they do *until* they reach home? Until Menzel's experiments, it had not been possible to track them in flight when they were out of sight out in the field. Menzel had a tool — radar but with a unique twist — whereby he could trace bees' actual flight paths over a kilometer away by radar and record them on a computer. And he invited me to come see the work in progress.

The problem of tracking small objects such as insects from a long distance by radar had always been that radar would "see" too much. You could not isolate and then plot a single specific bee out of all the extraneous noise of echoes bouncing off *all* objects. The new insect-tracking radar technique started in 1999, when Joe Riley, a British researcher, applied a radar system able to track very small objects over long distances by attaching to the insect a small device that, after receiving the energy of an electromagnetic sound pulse, would respond with a frequency *other* than that of the transmitted ultrasound. The receiver is then tuned to amplify only *that* frequency. In this way, it became possible to track the flight paths of individual preselected bees equipped with the appropriate transponders because the echoes from all other objects were filtered out.

The Menzel group's electronics technician, Uwe Greggers, adopted the Riley system in 1999 and 2001 and got interesting results, but then ran into software problems. Nevertheless, given the promise from the data they did get, the scientists contacted a radar specialist at Emden (north Germany) who agreed to develop the system. The Menzel group then needed to find the right site in which to use it. They needed to locate the experiments at a large flat area devoid of trees in order to be able to record the complete flight paths without interference such as the bees' trying to avoid objects or being attracted to them. The closest suitable area was an expanse of marshy meadow about a two-hour drive from Berlin. The large, idyllic farmstead near the village of Klein Lübben and land associated with it had accommodations for seven or more helpers, making this site amenable.

One Menzel group experiment in the works when I visited involved training individual bees to expect food at two widely separated feeding stations, but only one station at a time was open to them. I had no idea what to expect, and on my day with the team I was eager not only to watch the bees but also to see the experiment in action.

It was early in the morning when Menzel picked up Greg-

gers and me for our trip to the experiment site in the Branden-
burg countryside. We loaded a large, heavy printer that would be
used to handle the large-scale printouts of flight paths, and then
we were off down the Autobahn. Two hours later we arrived at
Klein Lübben, a quiet village of farmsteads that at least in outward appearance has changed little since medieval times. The
fields were several kilometers square, flat, and moist — perfect also
for frogs, and hence storks which nest there in baskets attached
to the tops of red-tiled house roofs. Swarms of starlings swirled
through the air, and a pair of white swans paddled serenely down a
canal along a dirt road, followed by a line of five still-downy gray
cygnets.

At one end of the study field stood a steadily turning radar apparatus with a large round antenna for sending out the signal. A
smaller dish antenna mounted directly above it would receive the
transformed signal bouncing off the transponder on an airborne
bee in the field. On the field sat two blue triangular tents and three
yellow ones. They were experimental landmarks for bees that could
be made available to them, to find out if they used them, and ma-
nipulated for experiments by changing their locations. In the distance sat a beehive, and I noticed a man running from it. He was
wildly slapping himself, in an obviously defensive mode. He had
been assigned to provide food for the bees close to the hive and then
was to gradually move the feeder into the field so that a population
of bees from that hive would be available for us to study when we
arrived at midmorning. He had come too close to the hive, and at
that moment it was he who was getting dispersed over the field, not
the bees. Also, as we soon found out, there were no bees coming to
the two feeder stations, as they were supposed to have been by now;
the student had apparently overslept or been otherwise distracted
from his assigned job of luring bees.

The experiment we wanted to do was in doubt. This was serious.

Two hundred thousand Euros had already been spent on this study, and the boss was intolerant of negligence. Luckily, bees from hives used previously for another experiment were still coming into the field to search for feeders. He could let some of those bees find the feeders and then train them to come back to specific locations.

For our experiment we needed to establish two feeding stations, A and B, separated by about three hundred meters. Certain bees were already keyed into the routine. When I walked across the field, one bee started following me. It looked most extraordinary: it had a lot of blue and green color, not just the usual plain brown honeybee attire. As soon as Menzel's helper and I set up our feeder, this specific bee landed on it and immediately started to suck up the rich sugar solution. Now I could examine her more closely: the green was a plastic tag with the number 29 on it that had been glued to her thorax. The blue was a slash of paint that had been daubed onto her abdomen.

Within a few minutes an assembly of several differently color-coded bees was lined up around the edge of the syrup dish. All were sucking up syrup. Some had green on the thorax, some had blue, and still others had yellow tags on their thoraxes, with additional daubs of white, blue, or yellow paint on their abdomens. Uwe Greggers and the unfortunate helper immediately started logging a list of the bees that had shown up in a notebook.

Each bee tanked up quickly, flew off directly toward her hive at the other end of the field, and then came right back to take a next load. Newly recruited (unmarked) individuals were also coming every minute to our site A. At the second feeder (site B) there was a similar flurry of activity, except it involved different individuals.

Menzel then instructed us to move our food station A one hundred meters closer to the second one, B. Bee numbers 29 and 30 green, both with blue tails, number 2 yellow with white tail, and number 39 green with red tail (who had all been present at A) then almost immediately started showing up at B, the new location. When crowds of bees had done the same, we removed one station and put the remaining one into the middle, between the two original sites. Next we moved our feeder to site B. Most of the bees, such as 30 green with blue and 39 green with red, who had been at the previous site, showed up. That is, we had trained bees who had been at one site to come to the second site, so we knew they now knew two sites and could potentially use either as a reference site to return home.

For the planned experiment it was important that the bees forget the intermediate sites that had been instrumental in getting them to go to the two widely separated sites. So, for the rest of the day, we alternately fed the bees, first at site A, then at site B, and monitored which individuals were showing up at *both* sites (most of the individuals continued to forage at either one or the other site).

We were now, near the end of the day, finally ready to move from training to trials. The experimental plan was to select one of the bees who knew both sites. This bee would, after feeding at one site and getting ready to leave, be captured in a dark box and thus "blindfolded" and then brought to a third feeding site where she had never been before. Here she would be released after being equipped with a radar-tracking transponder. We presumed that she would do at least one of three things: she might recognize where she was and fly straight home; she might instead fly off in her original (now wrong) direction; or she might immediately know that she was at an unknown location and search until she found one of her two feeding sites and from *there* take a direct beeline home. Knowing her exact flight path would allow us to distinguish among the alternatives, which would be essential to ultimately decoding her homing mechanism. Setting up this experiment had taken a long time, but I would now, possibly, be treated to an exciting demonstration of bee homing, one I could never have imagined possible.

Menzel picked up his walkie-talkie to call the radar station: "Mike, we're now going to put a transponder on a bee — are you ready?" Mike had spent some years in the army where he was trained on radar, and he was now working part-time while getting a university degree in computer technology. He replied yes, he was ready. Menzel then took me to feeding station A, where a whole lineup of bees was coming and going.

"Which one do you want?" Menzel asked me. I wanted a bee that I had gotten to know over the course of the day, so I chose 39 green with red-tipped abdomen. We waited for her to arrive and let her feed for a while. As planned, Menzel then held a glass vial over her while she was distracted sucking up syrup. When she was full, she walked up into the vial, and Menzel corked it shut and darkened it by wrapping his hand around it. We then took her to a site distant from both feeders, a place where she had not previously fed and from where we would now release her.

The vial holding "39 green" had a plunger at the bottom with a wide-mesh screen at the top. Menzel gently pushed the plunger in and forced the bee up against the screen, held her there, and picked up a tiny transponder (a wire holding a diode with a sticky pad at one end). With fine tweezers, he deftly removed the protective paper from the sticky pad and glued the transponder onto the top of the bee's thorax. "Ready?" he radioed Mike.

"OK."

Menzel removed the plunger and held the vial with the open end up, for the bee to crawl up. She hesitated at the lip of the glass, groomed her antennae, and then lifted off. She showed no strain in flight. (The transponder's weight is twenty milligrams, and a bee can fly with double her body weight, carrying a hundred-milligram load of nectar in her honey stomach plus two pollen packets on her hind legs.) However, she flew only two or three meters before dropping down into the grass, stopping to preen herself some more. But a couple of minutes later, she finally took off again. Mike, who was now monitoring her flight, radioed us. At intervals we heard: "She is heading south-north-east-north-west-south." Then, finally, Mike continued: "*Now* her path is straightening out — now she is heading *directly* for her hive!"

She had *suddenly* oriented correctly. This was the crucial point: she had apparently recognized something that had "placed her on the map," so that she then "knew" in what direction to fly to reach home. Assuming she had taken a path she had never taken before, did her successful homing suggest a "map sense"?

I ran over to the radar tent where Mike showed me the radar screen and the dots where the three-second successive readings traced the bee's path. A computer screen, where software had converted the time and directions of the bee's flight path into different-colored images for easy reading, showed that the bee's original flight direction was toward where the hive *would have been* had we *not* moved her from her feeding spot. In other words, she acted as though she didn't know where she was when we released her. As expected, however, after she reached the area where her hive would have been, she flew loops in several directions. Then, after she had flown ever-farther away from both her real and "would-have-been" hive locations, she suddenly seemed to orient and flew directly toward the hive. Amazingly, it was along a route that had *not* been her normal foraging route from her two feeding sites. Had she perhaps seen a blue or a yellow tent and, having learned their relationship to each other during previous orientation flights, transposed that information to fix her new location? Only more bees could tell.

Other bee homing experiments with hundreds of bees were ongoing. And in the group's final publication two years later, the thirteen-author research team headed by Menzel concluded that honeybees incorporate information for flight direction from both their previously learned flights as well as landmarks and from the flight directions learned from hive mates within the hive. But they can redirect their flight vectors to and from the hive and perform *novel* shortcut flights between the learned and the communicated vectors.

"The" homing instinct, recognized and traded on by every American beeliner to get honey, and used by von Frisch to decipher the bee language, is a source of fascination and mystery still. Von Frisch had likened it to a "magic well" from which the more you take, the more runs back in. The "well" is still doing that, three-quarters of a century after his prophetic pronouncement.

GETTING TO A GOOD PLACE

THE TENT CATERPILLAR MOTH, *MALACOSOMA AMERICANUM*, is common in North America. It emerges from its light yellow silk cocoon in late summer, and the female is then ready to deposit her batch of over a hundred eggs. She searches for an apple or a cherry tree, and somewhere out on a thin twig of just the right diameter — about a half centimeter — she exudes her eggs along with sticky foam to form her egg mass into a ring that wraps around the twig. The foam dries and hardens, encasing the clutch of eggs and gluing them solidly to the branch where they stay through the coming winter. But the larvae develop inside the eggs during the summer and, while confined in their eggs through the winter, hatch at almost precisely the day, about nine months after the egg-laying, when their tree breaks its buds.

The moth is named for the conspicuous communal homes of silk, called "tents," that its caterpillars make, and in the spring of 2013 I found a just-made tent on a young black cherry tree next to my Maine cabin. Like nearly everyone else in this part of the country, I was long familiar with these caterpillars but had not deemed them worthy of a closer look. The tents act, I learned, like miniature greenhouses and warm the new caterpillars at a time when nightly frosts are still common. But, despite its advantages, to have any home is to incur costs: it has to be made, and it takes time, energy, and expertise to make, and the wherewithal to travel to and from it. For the time being, I wanted to know where the caterpillars making this home had come from. To my surprise, the ring on the twig with the now-emptied eggs I was looking for was almost a meter from the tent. How had the many hatchling caterpillars "decided" or been able to stay together and then coordinate to make their tent? Squinting against the sun, I could see a glistening trail of fine silk leading from the emptied egg-case ring to their home, so here was at least a hint as to how they crawl together to end up at the same place.

On the second day after I found the tent, May 1, there was still snow on the ground in the woods. There was as yet no sign of fresh green anywhere. But I wrote in my journal, "Black cherry buds ready to pop leaves." These trees are the first to leaf out, and the caterpillars could not have fed yet. What would they do? An hour after the sun came up, the tiny caterpillars emerged from their tent and massed on its sunny side. An hour later they started milling about, and then a few started crawling, seemingly aimlessly, several centimeters up and down the trunk and branches of the cherry tree.

As I had anticipated, some of the tiny caterpillars started to crawl back onto the same branch they had come from, possibly following their previously made silk trail. But they went only six centimeters before turning back. Others went down the trunk of the tree. Always some would turn back, and then the others followed one behind the other in a line. Finally, by 7:30 a.m., a contingent of about twenty of them had progressed nine centimeters down the tree trunk, although two were coming back up. Then more started to leave the tent, and eventually all were in one long line, going only down the trunk and then angling up another branch. In half an hour the leaders had traveled seventy-three centime-

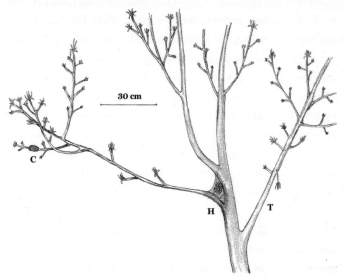

30 cm

C

H   T

*The young black cherry tree showing relative locations of a tent caterpil-
lar moth egg cluster (C) from which the clutch of just-hatched caterpillars
emerged and traveled to start making their home (H) in a crotch of the
tree, and their first travels as a group (T) to feeding places*

ters and reached a bud. The rest were strung out all the way to the
tent, but their two other travel-direction options had been abandoned. All were eventually massed at the same cherry bud, three-
quarters of a meter from their tent, and in an hour and a half they
had all returned to their tent, one following the other in a long
train.

At noon they came out and crawled onto the outside of their tent,
waving their heads back and forth, apparently weaving silk from
their salivary glands to enlarge it. Another hour later they were
again all massed inside the tent and perched, immobile, tightly
against the bark, where they were barely visible through the thin
gossamer veil of silk.

The caterpillars stayed in their tent through the night, and I expected them to go at sunup to the same branch where they had been the day before. But instead, this time they all followed an entirely different path, going directly up the tree instead of down as on the previous day, and without taking another side branch. I could not detect any silk on their so-far two different foraging trails, and this time they went even farther — a distance of 130 centimeters. After their one meal the day before, they were already noticeably larger. A few were the same size as the day before, but most had probably doubled in weight. There were many tiny fecal droplets in their web. So they had fed, even though it seemed hardly possible that they had anything to feed on at the barely opening bud.

On the third day the buds had opened and the tree was replete with new small leaves pushing out of the buds. But it had been a cool night — there was again frost on the ground at dawn — and the caterpillars made a slow start.

The pattern soon became clear: the caterpillars spent most of the night and most of the day when they were not feeding in their home. The time spent on tree branches was brief, and it could not have been just to keep warm that they stayed in their home because they went back inside just as quickly after feeding regardless of temperature or time of day.

Having found and watched the caterpillars of one tent, I then observed others for more clues to their homing behavior. One of the surprises to me was that as they grew larger, they foraged independently of one another, no longer going to and from feeding areas in groups. Furthermore, after they were about half grown they left their tents, not to return at all but still to continue feeding before eventually searching for a spot in which to spin their flimsy cocoon. Tent caterpillars usually choose a bark crevice to pupate, although commonly they also choose the cracks in the sides of buildings. But why were the young caterpillars strongly homebound and the older ones not?

I suspect the young ones' web-making behavior may have evolved in part as an anti-predator response. The tents were visited by red wood ants, *Formica rufa*, and right after the caterpillars hatched, these ants often loitered alongside them on their trails. I tore a nest open on one side to find out if it served as protection. It must have, because ants entered, though frequently wiping their antennae as though irritated. Nevertheless they tarried inside the damaged nest, and I saw them grab and walk off with caterpillars. No ants entered an intact nest of the several I watched, each of which consisted of several successive layers of silk. Thus, the webbing of the tent acts as a deterrent to predators such as ants. Staying inside the home most of the day and night, as these caterpillars appear to do when they are small, probably reduces mortality from parasitic flies and ichneumon wasps as well. When they are larger, the caterpillars are probably protected from the ants, as well as from most birds, by a layer of fine spines. They pupate without having to bury themselves to escape frost, because the adult emerges long before there is any frost.

Because these caterpillars are protected from predators in the summer homes they build and by the spines they wear, because they mature early enough in the summer for the pupae to avoid the cold of winter (by early emergence of the moth), and because the eggs and young larvae are immune to freezing because of the antifreeze they contain, "everything" in the life of the tent caterpillar moth may be found within a few meters. The adults that emerge in late June are not far from the apple or cherry trees where the parent left her eggs, and their life cycle can be completed without their having to go far from home, unlike some other insects which traverse a continent to be able to satisfy all their needs.

Monarchs. Of all the insects, the travels of the monarch butterfly, *Danaus plexippus,* are perhaps most famously spectacular in both scale and scope. Dr. Lincoln P. Brower of the University of Florida in Gainesville (now at Greenbriar College), who has studied this butterfly and its migration for over forty years, records the rich history of the emergence of our knowledge of monarch migrations. Early naturalists saw "immense swarms" in the prairie states where the caterpillars fed on the leaves of the many native species of milkweed (*Asclepias*) and the adults fed on the nectar of their flowers. Monarchs declined when later industrial agriculture destroyed many of their food plants, but in the nineteenth century they resurged in the East due to land clearing and the spread mainly of one milkweed, *A. syriaca.* Millions of them were seen passing for hours, even in Boston. This was a phenomenon that is hard to imagine now and it ignited much interest then. Charles Valentine Riley, the entomologist who first hypothesized that these butterflies engaged in a birdlike migration, cites people seeing them in the fall in swarms that extended for kilometers and obscured the sun, "blurring day into night." Huge lines of them passing Boston in 1880 were described as "almost beyond belief." Now, with reforestation, plowing, and then the use of Roundup and other weed killers that eliminated their food plants in agricultural fields, the monarch is but a shadow of what it was. In the past several years in the East, it seems to have almost disappeared. For the first time, I saw not a single one in late summer of 2013. But our knowledge of the scope of the monarch migration has blossomed.

Monarchs migrate on their own power for thousands of kilometers, and, unlike many other insect migrants, the population (though not the individuals) has a regular *two*-way migration, al-though as with the other insect migrants, the individuals that come back are not the same ones that left.

Unlike most of the other North American butterflies and moths, which overwinter in New England as eggs, larvae, pupae, or adults, monarchs cannot survive there through the winter in any stage. The population that normally now graces fields all along eastern North America overwinters at around three thousand meters' elevation in dense fir groves on the southwest slopes of volcanic mountains thousands of kilometers to the southwest, near Mexico City. The

The monarch butterfly adult, caterpillar, and chrysalis

monarchs find shelter in those fir stands from rain, hail, and occasional snow. It is not cold enough for the butterflies to freeze there, but it is cool enough for them to conserve the energy resources that they have accumulated on their way south.

In the summer, the monarchs fly in what look like random zigzag patterns over the New England fields as they stop here and there to sip nectar. Occasionally you see a mated pair, the female doing the work of flying, the male dangling passively with folded wings while attached by his genitals. After the prolonged mating (and/or technically "mate guarding," since it prevents mating by other males), the female glues her delicately patterned green eggs with gold markings, one at a time, to the undersides of milkweed plants. In a few days, the flashy yellow-black-white larvae hatch and start chomping. After about fifteen days (depending on the temperature), the caterpillars have increased their weight to 1.5 grams (2,780 times the hatchling weight). The caterpillar attaches itself to a support such as the underside of a leaf by a clasping organ at the hind end of its abdomen to hang upside down. It will then molt into the bright green pupa (chrysalis) with the shiny golden spots that is surely familiar to almost all school kids. In a few days, the chrysalis starts to turn dark, and the outlines of the orange-patterned wings are visible through the now-transparent cuticle. When the chrysalis splits, along a predetermined line of weakness in the back, the limp adult slips out and expands its wings, and in two or three hours hormones will have instigated a biochemical process that hardens its body armor and stiffens its wings. The butterfly is ready to fly. Where will its wings take it?

Thanks to the monarch studies initiated in 1935 by Dr. Fred A. Urquhart and his wife, Norah Urquhart, from the Zoology Department of the University of Toronto and continued to the present day with the input and cooperation of thousands of amateur volunteers, there is now an amazing story to tell. The Urquharts noted in the late 1930s that the monarchs they saw in late May and early June in Canada had tattered wings, and they knew that this species would not and could not overwinter in Canada, so they suspected that they may have come a very long way. Monarchs fly in a southwesterly direction in the fall, but nobody had a clue where they ended up. To get some idea of the butterflies' movements, these researchers in 1937 began gluing paper tags onto monarch wings with this message: "Please send to Zoology University Toronto Canada." Monarchs weigh almost half a gram and the wing tags only 0.01 gram, so the tags were not likely to hamper the animals' movements. Similar tags, used today, have pressure-adhesive backing and can be folded in half and glued over the leading edge of the forewing (after the scales are removed).

The idea from the inception of the monarch-marking studies was to try to find out if the butterflies migrated — an idea that at the time, as Urquhart noted, "was considered quite impossible." But the question of where the butterflies might be going to and coming from grabbed the imagination, and anyone seeing a tagged butterfly would be sure to try to catch it. Sure enough, tags were returned over decades that suggested a migratory pattern. Individual tags were returned from huge distances, up to 1,288 kilometers. One monarch that was tagged in Ontario in 1957 was recovered eighteen days later in Atlanta, Georgia, 1,184 air kilometers distant. Clearly, when the butterflies left Canada in the fall, they headed south.

Still, nobody knew what happened to the mass of butterflies. Then, in January 1975, Cathy and Ken Brugger of Mexico City found them — a dazzling, shimmering, orange display of an estimated 22.5 million monarch butterflies on one 2.2-hectare site (which turned out to be only one of ultimately thirteen overwintering sites in the mountains of Mexico). The millions of monarchs were festooned in the trees in the mountains of Michoacán near Mexico City. The Urquharts excitedly traveled to see the site and on

January 18, 1976, listened to "the sound of the fluttering of thousands of wings [that was] like that of a distant waterfall." As they stood awestruck by this dazzling display, a pine branch broke off from the sheer weight of butterflies attached to it, and it crashed to the ground right in front of them. Fred Urquhart had been posing for a *National Geographic* photographer surrounded by these just-fallen butterflies when, incredibly, he saw a tagged one among them. When he traced its origin, he learned that it had been tagged on September 6, 1975, by Jim Gilbert, from Chaska, Minnesota. Urquhart, who had encountered countless tagged butterflies in his career, said it was "the most exciting one I have ever experienced."

The picture that has now emerged from decades of study is that *individual* butterflies migrate all the way from Ontario to Mexico in the fall, arriving there at their overwintering sites in a torrent during October. They spend most of the winter in Mexico in a cooled low-energy state but soar around on warm days to drink water and replenish on nectar. In early spring, when their sex urge awakens, there is a mating orgy followed by a mass exodus. Most of the females mate before leaving, and their "compasses," which were set to take them south in late fall, are now "reset" to take them in a northerly direction.

As the tide of butterflies advances northward, the females stop to lay their eggs on milkweed. Some of the butterflies from Mexico make it all the way to the north, and others (their offspring) that grow from the eggs laid along the way arrive later. Those of the first generation have slightly tattered wings when they arrive in the north, while those that arrive later have untattered wings. (However, not all monarch populations migrate, and not all that do, travel in the same directions as the populations of northeastern North America.)

One of the mysteries that puzzled Fred Urquhart was *how* the butterflies home. In Urquhart's 1987 book on the monarch, he speculated that the butterflies perhaps use the Earth's magnetic lines of force, although different populations of the butterfly migrate in different directions, so they could not all be orienting to it in the same way.

A potentially even more puzzling question is the ultimate (evolutionary) one of why these butterflies migrate in the first place. Urquhart simply suggested what he admitted was a "perhaps farfetched" idea: that "twice each year it [Earth] passes through an area rich in some sort of radiation that could impinge upon animal life [that] might affect in some manner the cells of the body causing reproductive organs to abort in the fall and develop in the spring and initiate the migratory response." This is an unlikely theory, though, mostly because it depends on a mechanism that is not adaptive in evolutionary terms. Instead, more current thinking about the adaptive reason *why* the phenomenon has evolved focuses on energy economy and maximization of resource use under the expected evolutionary constraints from the monarch's having evolved in the tropics, meaning it was not able to survive northern winters. (Monarchs belong to the family Danaidae, an otherwise strictly tropical group.) Migration to the north in the spring opens up the milkweed crop over a major swath of North America as a food base for the larvae. In addition, the journey is probably not costly to the monarchs, either in terms of predation (since they are chemically protected from predation by poisons they sequester from their food plants) or in terms of energy costs, since their energy intake along the way more than makes up for the energy expended for travel. Indeed, unlike most birds that may deplete all their fat reserves on migration, these butterflies instead fatten up on their journey and may consist of about 50 percent body fat by the time they arrive in Mexico, where their overwintering fast begins.

Butterflies and moths experience tremendous selective pressure, and undoubtedly there are constant readjustments of survival strategies. Weather affects the populations, not only through flight activity and flight range as well as growth rates of larvae, but perhaps also indirectly by influencing virus infections. But Urquhart noted that each female monarch butterfly lays up to seven hundred eggs, and he calculated that the "biotic potential" — the number of individuals if there are no deaths — of one female after only four generations (that is, at the end of one summer) is 30,012,500,000 adults. Luckily for the planet, animals' reproductive potentials are never naturally realized, for long. The limit is quickly reached when the population uses up its food base, in this case milkweed. In some years a virus decimates most of the monarch population over North America, but then several years later it rebounds. But the population cannot rebound from some things: in recent years there have been massive declines of the monarch population that cannot be reversed, because they are due to unnatural causes — the massive conversion of land to crops, and the introduction of genetically modified crops that tolerate herbicides, which have allowed the elimination of milkweed that formerly grew between rows of corn.

The flight performance of monarchs is spectacular, but like the hordes of cluster flies from the surrounding fields and woods that overwinter in my cabin, they are traveling to a specific place for overwintering where they have never been before. Such homing movements are diverse, but common. Robert D. Stevenson and William A. Haber of the University of Massachusetts, Boston, found a regular seasonal migration of about eighty percent (250 species) of butterflies living in the dry lowlands of the Pacific Slope of Costa Rica that migrate to wetter forests of the east. Distances traveled range from ten to a hundred kilometers.

In North America as well as in Europe, the cosmopolitan painted lady, *Vanessa cardui*, a mostly orange and black butterfly with white spots and pink and blue "eyes" on its under-wings, at times appears

Red admiral butterfly larva, adult, and chrysalis. The larva makes a shelter for itself by pulling leaves together and holding them with silk, while then feeding on the leaf.

in large numbers and then is not seen again for years. Usually the individuals are seen crossing a road, and almost all will be heading in the same direction. The painted lady regularly migrates north from Mexico, from where it originates, after heavy rains in the deserts have created an abundance of food plants, primarily thistles. A friend told me of one migration while he was in Arizona when his windshield wipers "soon became useless" because of the huge numbers of painted ladies plastered onto them as he was driving. I see them regularly in Vermont and Maine, but seldom in large numbers (the summer of 2012 was one of the exceptions).

One of the butterflies that not only migrates as an adult but also hibernates in some parts of its range is the red admiral, *Vanessa atalanta*. It is (as are all butterflies!) beautifully colored. It sports a wide red stripe across each dark forewing ornamented with white spots, and its larvae feed on nettles. I wrote in my journal on May 11, 1985, near my home in Vermont: "In the afternoon from around 2:30 to 4:30 PM, as I was jogging along on an 18-mile circular loop I counted 512 red admiral, crossing the road in front of me. All but 5 of these were flying in a northeasterly direction. At 5:00 PM, after I was home, I take compass readings of butterflies flying over a plowed field where they funnel onto it through a valley. I can see them to take a bearing for at least 50 paces — 250 feet. All 22 that I observed flew in NE direction. At 6:00 PM activity almost stopped. The breeze is slight, from northwest." In the summer of 2001 and again in the spring of 2010 I saw large numbers of red admirals. They fed on freshly opened apple blossoms, and later all the nettle plants in a neighbors' sheep pasture had an abundance of their caterpillars.

Moth migrations are perhaps more spectacular than those of butterflies. Jason W. Chapman and colleagues report one recent ten-year study involving radar tracking of about one hundred thousand owlet (Noctuid) moths, primarily the silver Y moth, *Autographa gamma*, migrating south in the fall from northern Europe, and then north from the Mediterranean in the spring. Like the butterflies, these insects breed along their migration route. Also like the butterflies, the moths partially correct for crosswinds, to maintain specific directions. Most surprising perhaps is the moths' windsurfing; they choose the most favorable wind currents corresponding to their respective spring or fall migratory directions. If the wind shifts about twenty degrees from the favorable direction, they adjust their flight to accommodate and maintain the correct direction. If the wind shifts ninety degrees, though, they stop and wait for a favorable wind. Millions of them fly together in the dark of night, and, like the monarchs', their compass directions are likely tuned to the Earth's magnetic fields. Some studies of radio-tagged green darner dragonflies, *Anax junius*, suggest that these insects also migrate hundreds to thousands of kilometers from north to south with those that return being a different generation.

These behaviors get the animals to a good place (for overwintering or for reproduction). Like the long-range movements with specific endpoints on the map, homing to a good place is not always easily distinguished from moving out of a bad place. The behavior is a mechanism with deep evolutionary roots. Indeed, insect wings (and metamorphosis) may themselves have been an original adaptation for dispersal, to colonize temporary pools, animal carcasses, or other temporary resources. The first individuals to reach the resource won the competition to use it and multiply there, and these were more likely to be the ones that flew, and flew far and wide, rather than those that walked at random.

Wings and metamorphosis have lesser value in constant conditions. Some insects are able to respond in real time to the changes in conditions they experience (especially crowding), in that when they don't "need" to disperse they either don't grow wings (some aphids) or the muscles to power the wings are broken down and the amino acids from the protein are used instead to make more eggs (in some Hemiptera bugs). Often there are discrete "dispersers" versus "non-dispersers" in any given insect population, and the percentage of each depends on the quality of the home habitat and hence the relative cost/benefit ratio of moving versus staying.

Dispersing to "anywhere but here" generally applies to nonmigratory species that have no encoded or learned directions to go to but may have innate instructions to move in more-or-less straight lines rather than potentially going in circles in order to achieve distance. In Africa, dung-ball-rolling scarab beetles race away from their often thousands of competitors at a dung pile at night by using the swath of stars of the Milky Way galaxy as a reference. Swarms of insects feeding at dung and carcasses also attract predators, and as soon as they finish feeding, many distance themselves from those predators. I've observed blowfly larvae at animal carcasses keeping to almost perfectly straight lines in their getaway at dawn, by steering directly toward the direction of the rising sun. Mass movements sometimes observed in some rodents, such as lemmings and gray squirrels (as in 1935 in New England) following a population explosion after a superabundance of food, may be another example of dispersal to get to a better place, though not necessarily a predetermined one.

On the other hand, "true" migrants are able to utilize ideal conditions in two places, provided they vary predictably. Arctic terns, *Sterna paradisaea,* breed throughout the Arctic, then fly to Antarctica to escape winter when food availability declines and to arrive in spring and food again, a round-trip distance of nearly seventy-one thousand kilometers. Gray whales, *Eschrichtius robustus,* also feed in the Arctic in the summer but then travel eight thousand kilometers along the coastline to Mexico to bear their calves in warm waters.

Dispersers are not neatly differentiated from migrants, although the first commonly rely on passive mechanisms as opposed to the migrants, which move to specific goals by their own powers of locomotion. There are all gradations in between. Each case is unique, and there are thousands. Let's look at more ways of getting to a good place, as represented by eels, a grasshopper, aphids, and ladybird beetles.

Eels. There are many species of eels, but the American, *Anguilla rostrata,* and the European, *A. anguilla,* the species with which we are most familiar, live most of their lives in freshwater ponds and lakes. For reasons that are not clear, though, they do not reproduce in their home areas. To the contrary, both disperse (or "migrate") thousands of kilometers on a one-way trip to spawn and then die in the mid-Atlantic.

Just as birds and some insects use air currents, eels use water currents to help them leave their lifetime homes. After eels leave their freshwater homes and head for the ocean to spawn and die, their larvae then drift in the ocean currents for years. But eels' dispersal behavior is anything but passive. Eels fatten up to prepare to leave their home haunts in the bottoms of freshwater lakes and streams. Before they head for the sea, they absorb their digestive tracts and transform themselves by greatly enlarging their eyes and turning silvery on their ventral side. The latter transformation produces counter-shading that reduces their visibility to predators below them in the open ocean waters.

The eels' dramatic changes in behavior, morphology, and physiology that enable them to switch from living in freshwater habitat to open ocean highlight the operation of strong selective pressures. But why do they leave their homes in freshwater ponds where they grew up and lived most of their lives? Their one-way, once-in-a-lifetime migrations to their Sargasso Sea spawning grounds can't be to find a better feeding ground, or to escape competition. However, I suspect what the behavior accomplishes superbly is that the adults, which are predators, do not come in contact with and feed on their own young. Is it a mechanism that has evolved because it reduced predation on themselves?

There are over six thousand publications on eels, but the life cycle of these economically important food fish is still murky and has a long history of speculation. For centuries, nobody ever saw a baby eel, and even now, their spawning has not been witnessed. Aristotle presumed that eels grew from earthworms. The first young of eels, transparent leaflike forms, were found in the open Atlantic

Ocean. Gradually, as more of these leaflike creatures were collected, it was noted that they varied in size, and that the smallest ones were found south of Bermuda in the Sargasso Sea, which was therefore presumed to be the site of their origin, that is, the eel spawning area.

The eel larvae, after hatching in the waters of the Gulf Stream, drift north. Like plankton, they move at the whim of the prevailing oceanic current. As they grow from about five to six centimeters in a year, they take on a more eel-like form but remain transparent. They are by then able to swim and, presumably by scent, find and swim up a river. Unlike salmon, however, these transparent "glass eels," as they are known at this larval stage, can have no specific home stream scent to follow, because they have experienced only the scents of the ocean.

The female glass eels at this stage, in early spring, migrate up rivers and streams along the East Coast of America. After two months in a river they grow to about ten centimeters. Now known as "elvers," they are no longer transparent, and they enter lakes and become eels. These female eels live in lakes for eight years or more, fattening up (the males stay in the saline estuaries). When they achieve the right amount of fat, these females become sexually mature. Each develops a clutch of three to six million eggs, and then one fall the gravid females start their journey downstream back to the ocean, to the Sargasso Sea to spawn. Since the males don't live in fresh water, somewhere in the ocean the females then apparently meet the males for fertilization.

As the Gulf Stream continues north beyond North America, the larvae of the European eel, which originate from the *same* apparent spawning area, the Sargasso Sea, as the American eels, continue their journey. Finally, in two or three years, they reach the coasts of Europe, where they then also seek rivers and streams. Migrating upstream, they become pigmented, and after growing to adulthood, they migrate back into the open Atlantic and make their sixty-five-hundred-kilometer return to the Sargasso Sea, where they spawn on average several million eggs, and then die. Only one of several million of them will make the return journey to grow to a reproducing adult in fresh water.

Grasshoppers. One of the best-known insect dispersers, the "migratory locust" (the grasshopper, *Schistocerca gregaria*), engages in some of the most spectacular mass movements in the animal kingdom. On the African continent, this species has been famous since biblical times. Swarms of the "locust"

Nymphs and adults of the two phases of the migratory grasshopper

have blackened the skies, and as those in the vanguard settle onto the earth and consume every green thing where they land, the rest fly over them until they reach more green, while those behind then take flight and do the same, and so a horde of hundreds of millions moves along, stripping all vegetation in its path. Predators cannot put a dent in those hordes. Additionally, the migratory locust is distasteful to potential predators because when it migrates, it is not choosy about what it ingests and takes up toxins from poisonous plants, which it incorporates into its tissues. The grasshoppers' bright red-orange and yellow coloration, like that of many insects including the monarch butterfly, reminds potential predators of its distastefulness.

Although this distinctively colored grasshopper appears to arrive suddenly, it is often there all along, but in a different guise. It has a green solitary grass-fed form that blends in with its food and that is palatable to predators. For a long time scientists thought that the grasshopper "migrants" that appeared so suddenly were a unique species, one arriving from an unknown origin and heading for an unknown destination. Now we know that the migrants are a "phase" of a common species that changes its color, form, and behavior in response to crowding. Proof comes from experiments: to create these "migrants" from isolated individuals one takes a nymph (immature stage), puts it in a jar, and has a motor-driven brush tickle it continuously. The constant tickling mimics the crowding, which in the case of *S. gregaria* is the signal evolution has "chosen" to trigger the nervous system to alter the hormones that result in development into the restless migratory phase of a different color, wing length, and behavior. It is a good example that shows environment is "everything," or from another perspective, it's all about genetics.

Migratory-phase locusts are highly irritable and will jump up

and follow a crowd flying over it. This behavior removes the grasshoppers from an area that is overpopulated and brings them to new
land where conditions are conducive to feeding, egg-laying, and
growth of their offspring. Although the grasshoppers could have no
knowledge of where such a distant but good place might be, they
migrate to it as if they do.

The grasshoppers reach a consensus. It is a sensible one, although it involves no thinking and no discussion. They simply fly
up to join the crowd, which follows the prevailing winds. Eventually
these winds meet air from an opposite direction and, when moist
tropical air rises into cooler altitudes, rain precipitates out of the
resulting clouds, depositing the grasshoppers to earth along with
the rain. As the ground is watered and softened, the grasshoppers
can shove their abdomens into the soil to lay their eggs. The new
nymphs hatch just as new food starts to sprout. Their homing (or
"dispersal"?), which has ended at this good place for them to reproduce, is now complete.

Aphids live in crowded "colonies" on plants into which they
insert their mouthparts, much as mosquitoes puncture skin,
except that they imbibe plant sap instead of blood and may
stay plugged in at the same spot for most of their life spans.
One might suppose they could not or would not migrate. But,
like the migratory grasshoppers, they may travel possibly
hundreds of kilometers. Nobody knows for sure how far; it
depends on the prevailing winds.

Sedentary aphids already on good food do not leave to seek, or
even require, mates. Instead, they switch to virgin births after a sexual migratory phase. Daughters then settle directly next to mother,
and so on and on for many generations as the colony grows. And
then, cued by the shortening of the days in the late summer and fall

when the food supply runs out, the aphids' offspring take a different developmental route. Because of changing day length and/or food, the nymphs on their final molt grow wings and become sexual. Frail and weak-winged they are, but an aphid is light and carried by the wind much like the seed of a dandelion or poplar tree, or a baby spider on a thread of silk. I usually see them in September when they appear like flecks of white lint floating erratically in the air.

To reach wind the aphids fly or are wafted *up*. Eventually, they don't fight the wind but drift along and settle somewhere back down to earth. On their descent, assisted by their wings, they head toward anything colored light green. This color (unless they are tricked by pieces of green paper coated with sticky glue left by an insect physiologist studying them) is likely to be associated with their favorite food, fresh plant growth. After landing, perhaps because the chances of a mate arriving at precisely this one tiny spot of residence are remote, they switch back to virgin births and thus restart the cycle.

Ladybird beetles, the predators of aphids, similarly have adapted by migrating in a seasonal environment. In the western United States they migrate mainly on their own power from lowlands up into the Sierras, where they can in some locations be scooped up by the bucketful (generally to be sold to farmers and gardeners — to control aphids!). At the campus of the University of California at Berkeley, I often saw streams of them flying or being blown uphill in Strawberry Canyon by the campus when the grass was drying after the spring rains.

Ladybird beetles of some species migrate when reproduction must cease for the season. Despite the energy they expend for flight, they may migrate largely to *save* energy. It goes this way: As long as there are plenty of aphids to be had, both larvae and adult ladybirds don't go hungry. Eventually, however, the green vegetation suitable for aphids disappears in the hot California summer, and so the aphids leave. Now the beetles' resting metabolism kicks in as a significant liability. Resting metabolism for beetles is high at high body temperature but becomes almost negligible when they are torpid at the lowest body temperature tolerated, near or slightly below freezing. An elevated resting metabolism, month after month in the western states' dry hot summer, would deplete both the beetles' energy reserves and their body water. Without replenishment of food and water in that environment they would die. But by flying, with the aid of thermals, they are brought into upwelling air currents in the hills and then into the mountains where they reach cooler air. At this point, though, they do something different from the aphids: instead of being attracted to green, they are attracted to either red and/or the scent of each other. How else to explain that they crowd together into large groups in which they then overwinter? The advantage of their grouping behavior is not clear, but I suspect that it amplifies their noxiousness. (This is based on experience: ladybird beetles regularly come into my cabin to overwinter and quite often crawl into bed with me. I can vouch for the fact that they are noxious if not obnoxious, and several more so than one.)

The ladybird beetles arrive at a suitable place — a cool one — where they conserve their limited energy reserves during winter. The hypothesis that they home not just to an area, but also to a specific spot, is based on observations that my friend and colleague Dr. Timothy Otter has made in the Sawtooth Mountains near Stanley in Idaho. The beetles there were known by local ranchers to aggregate every fall in large numbers in specific rock cairns of decomposing granite in the hills above the valley floor. Otter, a biologist, wondered why the beetles aggregated there, in specific spots, and not in similar places nearby.

Since hibernation concerns adaptations related to temperature,

Otter concentrated his efforts on unraveling the beetles' temperature tolerance and compared it to the temperature at that site. But, curiously, there was nothing unique about the temperature of the specific site on "Ladybug Hill," relative to other sites near it. Nor did the grouping of the beetles affect their temperature; the temperature of their aggregation was indistinguishable from ambient temperature. So, they didn't aggregate to keep warmer than the environment there.

As already mentioned, the one thing all ladybug beetles do have in common is that they stink. The evolutionary significance of their aggregations is therefore likely the conventional one of other animals, namely, to advertise their noxiousness for protection from predators. Shrews and other predators may kill two or three victims and then spit them out, but they then learn to avoid them and so they do not continue eating the hundreds of thousands they would consume if they had found a bonanza of ladybirds. Given their safety of numbers, it is advantageous for any one beetle to join a group rather than to overwinter alone, because the chances that *it* will become a victim of a shrew's educational process are reduced in more or less direct proportion to the group number. This rationale for large numbers of beetles, but of different individuals, massing repeatedly at the same location is not proven, but my colleague Daniel F. Vogt and I found that it applied to another noxious-smelling beetle, the whirligig (Gyrinidae) water beetles in Lake Itasca in Minnesota. These beetles there homed in on any existing groups of tens of thousands at dawn, after a night of foraging far and wide on the surface of the lake.

The second question is: *How* is an aggregation formed?

Insects have an impressive ability to home in on scent, and ladybird beetles could find the aggregation by following an odor plume. Memory cannot be involved in the ladybugs that Otter found year after year aggregating at the same site, which were generations removed from those of the previous year. The beetles arrive on site in September, stay there eight months, and in May return to the valley floor to feed, mate, and reproduce. Only their descendants, two or three generations later, could return to the tiny spot where they had hibernated.

The aphids' rule of flying up toward the light, to then be dispersed by the wind, and then homing in on green when they settle could be a model of what happens in ladybird aggregation. Ladybirds are much stronger fliers than aphids, although they too are swept along in updrafts. But such drifting, while helping to account for their annual ascent from the lowlands into the mountains, does nothing to explain how tens of thousands of them end up under the *same* rock pile.

If the ladybirds home in on color, this could be tested, as the aphids' homing in on food plants was tested — by leaving color targets at sites other than the traditional hibernaculum. But even if red color is an attractant (highly unlikely because the beetles aggregate under the rocks, not on them), that still would not explain their annual return to the *same* place in successive years. Do they smell their ancestors? Could thousands of smelly beetles piled up for eight months leave sufficient scent residue to serve as a marker that allows others to home in on the spot? If so, this idea could also be tested, by transferring an aggregation of beetles to overwinter at another physically similar place in the same general area as the old to see if a new traditional homing site for their mutual protection is created.

BY THE SUN, STARS, AND
MAGNETIC COMPASS

*Life has unfathomable secrets. Human knowledge will
be erased from the world's archives before we possess the
last word that a gnat has to say to us.*

— Jean-Henri Fabre

CHARLES DARWIN REFERRED TO THE ACCOUNT OF FERDI-
nand von Wrangel's Arctic explorations, *The Expedition to North
Siberia,* concerning how we home, quoting von Wrangel on how
the Siberians oriented by "a sort of 'dead reckoning' which is
chiefly affected by eyesight, but partly, perhaps, by the sense of
muscular movement, in the same manner that a man with his eyes
closed can proceed (and some men much better than others) for
a short distance in a nearly straight line, or turn at right angles
and back again." Darwin compared a bird's homing capability with
that of people, but much less favorably, by telling how John James
Audubon kept a wild pinioned goose in confinement, which at migration time became "extremely restless, like all other migratory
birds under such circumstances; and at last it escaped. The poor
creature immediately began its long journey on foot, but its sense

of direction seemed to have been perverted, for instead of traveling due southward, it proceeded in exactly the wrong direction, due northward." I'm not the least surprised at the behavior of the goose but all the more puzzled by our own orienting, which involves knowledge versus a feeling of "sense of direction." I recall instances of waking up in "total" dark, "knowing" in my mind precisely how I am oriented relative to the room and hence the rest of my environment, but irked by "feeling" that I am in the precise opposite direction. It is then a struggle to get the two to agree, which happens only after some effort.

In Darwin's time it was still supposed that humans had overall superiority over other animals. His then-hypothesis (later theory, and now fact) of evolution, which now binds us all as kin, was still revolutionary. Darwin found the goose's behavior puzzling because he could not know that geese, cranes, and swans stay together in family and larger groups and that although the young by themselves do not know the correct migration route, they learn to know it from their parents which in turn learned it from theirs. Other birds have their migratory directions genetically coded, and they go strictly by "feeling," since many of these have no knowledge because they migrate ahead of their parents.

We humans get lost easily. We would not get far without reference to landmarks, and I base that conjecture on (inadvertent) experiments. In one I was in long-familiar woods and got caught in a heavy snowstorm. Suddenly I got "turned around," and it seemed as if all landmarks had in almost one instant been erased. But I kept going, trying to maintain a straight line by trusting my "sense of direction." When I thought I had reached a place that I knew, where I should be going downhill, the landscape was instead sloping upward, and the brook I had expected to see was going in the "wrong" direction. At that point, knowing I was lost and no longer referencing to any signal, I backtracked in the snow and discovered that I had been walking in a circle, all the while thinking I was heading in the "right" direction. Yes, we can walk, a short way, in a relatively straight line with our eyes closed, by a process dubbed "path integration," but my emphasis here is on "relatively." Mice may do better. A friend told me of catching a Mexican jumping mouse in a live trap baited with peanuts. It had a kink in its tail, so he called it Crooked Tail, or CT for short. After he had released it several times, and it always returned for another snack of peanuts at the same source, he finally decided to "test its mettle" and released it exactly one kilometer away in thick brush and grass. The next morning CT was back for another snack. After release from two kilometers, though, it did not return. We don't know, though, if this was due to failed navigation, finding a new food, a cost/benefit calculation, or a run-in with a coyote, owl, or weasel. On the other hand, when I failed to navigate, I was positive that I had been going in the precise opposite direction, which meant I had no sense of direction whatsoever, except that coming from visual landmarks from which I had constructed a map in my head.

When we do home, it is by maintaining a constantly updated calculation from at least two reference points, and the motivation to use them. We are innate homebodies, normally seldom displaced, so that in our evolutionary history there has been little need for a highly developed home-orienting mechanism. Simply paying attention to familiar landmarks suffices. Males on average may perform better than females in negotiating unknown territory, and it is posited that they, having been hunters traditionally, have a better "sense of direction." But I doubt it. Learning, and especially attention, is hugely important for a presumed directional sense that can be developed to a high degree, as shown in some Polynesian seafarers living on isolated islands. But basically that involves being alert to more cues. These seafarers had been trained from near infancy to "read" the stars, the ocean waves, the winds, and other signs so that they may navigate over vast stretches of open ocean. But what a select few human navigators can accomplish with experience and with tools, many insects and birds do routinely as a matter of course, and with far greater precision over distances that span the globe.

Every fall and spring billions of birds travel to their wintering grounds where they can find food, and in the spring they return to near where they were born in order to nest. In huge tides, partially aided by favorable winds but mostly by their own muscle power, they ply the skies in the day and at night in the Northern and Southern hemispheres, sometimes covering thousands of kilometers in a few days. For the most part, the birds have pinpoint home destinations, places such as a specific woodlot, field, or hedge. In the fall they reverse their journey, though often by a different route, again to reach specific pinpoints in their winter homes. Turtles on the seas accomplish the same navigation feats between breeding and feeding areas.

The magnitude of birds' migratory performances staggers our imagination, in terms of both physical exertion and feats of navigation, because they are vastly superior to anything we could, as individuals, accomplish. Bird migration, as we now understand it, for centuries seemed impossible because we used ourselves as the standard, and that of turtles was not even considered. The animals' performances would still seem impossible, given our ignorance and arrogance, were it not for the proof from countless research experiments.

The homing behavior of birds was known and used as early as 218 BC, when Roman foot soldiers captured swallows nesting at military headquarters and took them with them on their campaigns. They put threads on the swallows' legs with various numbers of

Westminster City Council
Queens Park Library

Borrowed Items 10/08/2017 16:41
XXXXXXXXXX6611

Item Title	Due Date
* The girl in the spider's web	31/08/2017
* The diet myth : the real science behind what we eat	31/08/2017
* The homing instinct : meaning and mystery in animal migration	31/08/2017
* Cast iron	31/08/2017

Amount Outstanding: £0.05

* Indicates items borrowed today

Looking to fill your new
e-reader? Download e-books and
e-audiobooks for FREE at
www.westminster.gov.uk/libraries

knots to specify perhaps some prearranged signal or information, so that the marked bird when released and then recaptured at its home nest would bring the message. Today, between 1.1 and 1.2 million birds are banded annually in America alone, providing an ever more detailed picture of where the different species travel and when.

As with insect dispersals/migrations, our attention and insights into bird homing were and still are stimulated by spectacular examples. We are perhaps most impressed, if not baffled, not only by the birds' wondrous physical capacities, but also by the cognitive or mental capacities that underlie them. Seafaring animals, like albatrosses and shearwaters and sea turtles, are especially noteworthy to us because we can't explain their behavior by the use of at least to us visible landmarks, our main if not only recourse.

The Manx shearwater, *Puffinus puffinus*, navigating over the vast oceans, was one of the first birds to excite our curiosity enough to spark examining the wonder of bird homing. Shearwaters never cross land. All their food is taken from the water surface. As with most birds, their young are fixed to a specific safe or sheltered place, in this case an island, where one parent may spend as much as twelve days at a time ceaselessly incubating before being relieved by its mate. They nest in a burrow in the ground on islands in the North Atlantic, making it quite easy to catch, mark, and release them to identify individuals. We can also assume that as with bees, their motivation is to return home, and thus they are ideal subjects for homing experiments.

Prior to the First World War, the English ornithologists G.V.T. Matthews and R. M. Lockley took two shearwaters from their nest burrows on the island of Skokholm off the southwest coast of Wales and released them from points unknown to the birds. Under sunny conditions, the shearwaters returned to their nests by flying directly in their homeward direction. In one such test, a shearwater was carried by aircraft to Venice — a huge distance from its nest and an area where no shearwaters occur. The released sea bird might have been expected to fly south to the sea. Instead, it headed directly northwest to the Italian Alps and in the home direction toward Wales, in a path it never would have flown before. It returned to its home burrow on Skokholm 341 hours and 10 minutes later. This could, of course, not have been a direct nonstop flight. Unfortunately at that time there was no way of knowing if it had stopped to forage and/or what route it had taken.

The experiment was repeated involving even greater distances, after transatlantic plane travel became routine. Two banded Manx shearwaters also taken from Skokholm were carried by train to London in a closed box and flown to Boston, Massachusetts, on a commercial TWA flight. This is perhaps the ultimate in terms of the "blindfolded" displacement that I previously described for experiments with honeybees. One of the birds did not survive the journey to America, but the other, which was released near a pier on Boston Harbor, "abruptly turned eastward over the ocean." Dr. Matthews, a leader in the study of bird homing at the time who had released 338 Manx shearwaters on the British mainland, discovered the bird back in its home burrow before dawn on June 16, twelve days and twelve hours after it had left Boston, almost five thousand kilometers away. On reading its tag, Matthews sent a telegram to the person who had released the bird: "No. Ax6587 back 0130 BST 16th stop-FANTASTIC-MATTHEWS." Making another round that night to check on the bird again, Matthews, as though not believing his eyes the first time, wrote in a letter (to a friend, Rosario Mazzeo) that he was "completely flabbergasted" and had to read the ring several times before putting the bird back into its burrow.

By 1994 biologists had attached radio transmitters to animals that sent out high-frequency radio pulses received by satellites orbiting up to four thousand kilometers away. When two satellites picked up the same signal, scientists could calculate the transmitter location and relay it to receiving/interpreting sites on the ground. There, computers tracked the birds' positions and drew maps of their travel routes over months. From these and other studies, we have learned that these seafarers, and sea crossers, both turtles and birds, may wander over thousands of kilometers of the ocean vastness and then return to tiny isolated targets, the homes where they were born. They can travel in straight lines even at night and while correcting for the drift of currents or wind. Using the new technology, these behaviors have been demonstrated perhaps surprisingly in a sandpiper, the bar-tailed godwit, *Limosa lapponica baueri.*

The bar-tailed godwit, a shorebird that nests on the Arctic tundra, winters in the far south of Australia. It has a long thin bill for extracting worms from deep soft mud. This species makes its Arctic home on a shrubby hillside with low tundra vegetation and nests there on almost any of millions of hummocks to be found on the tundra in Alaska or Siberia. Its nest is a slight depression lined with grass and lichens. The female lays her clutch of four large olive-brown mottled eggs into it, and the pair take turns incubating for about a month until the fluffy young, in camouflage down, are hatched. The parents then lead their chicks around and they feed themselves.

The bar-tailed godwit is not a particularly unique shorebird, as such. (The Hudsonian godwit, *Limosa haemastica,* performs similar flights from Manitoba to Tierra del Fuego and back.) But in the past ten years, possible extremes of homing ability and some astounding physical capacities that back up this behavior have been revealed by Robert Gill Jr., a biologist with the U.S. Geological Survey, who deployed twenty-three godwits with either solar-powered backpack transmitters or battery-powered surgically implanted ones in the abdominal cavity. The transmitters trailed thin antennas behind the birds, and the radio signals from them indicated

Flock of bar-tailed godwits on migration

their location and were received by polar-orbiting satellites. The data of the godwits' locations throughout their flights was then calculated on the ground. Nine of the transmitters functioned for two years, yielding data on both the southern fall migration to Australia as well as the spring migration back home to the breeding grounds in Alaska.

They revealed the hugely surprising fact that the godwits make the flight from Alaska to Australia nonstop.

The godwits fly directly across the Pacific Ocean in six to nine days. One female covered 11,680 kilometers in 8.1 days in her southward migration, and another traveled 9,621 kilometers before she lost her transmitter after 6.5 days. When the birds arrive back in New Zealand or Australia after their transoceanic flight — with no feeding, no drinking, and presumably no sleep — they have halved their starting body weight.

. . .

Portrait of a bar-tailed godwit

The godwits' *northward* journey to the breeding grounds may involve a different route, and this one includes stopovers on the way. These stopovers permit the birds to replenish so they don't arrive emaciated just when they begin the most energy-demanding part of their breeding cycle. For example, one godwit, identified as "E7" (which covered twenty-nine thousand kilometers in a round trip from New Zealand to its nesting area in Alaska), on its northward journey stopped at several staging (refueling) sites in the western Pacific and Japan, from where it then made the relatively short jump to its western Alaska home. On the other hand, on its southern migration after the nesting, it flew directly south from Alaska across the Pacific and back to New Zealand.

Right after a male godwit arrives back at its patch of tundra that is its home in Alaska, he circles for hours high in the sky and calls loudly near this chosen home site. In as little as a week before, he may have been on a coastal mudflat in Japan, where he had a raging appetite and gobbled worms and crabs day and night. Similarly, to prepare for his departure before the Alaska winter freeze-up in the fall, he will feed until he has doubled and even almost tripled his body weight in fat. And then, by our standards, in grossly obese condition, he lifts off to fly south. Although some godwits will stop off briefly in the Solomon Islands and New Guinea, others will fly up to fifteen hundred kilometers per day without a single stop. On their stupendous flight the godwits use up not only their body fat but also protein derived from shrinking muscles and organs, including almost every part of the body except the brain. The flight muscles are the primary powerhouse for the effort, but the brain — the organ that drives birds' motivation to keep going — is more important.

Why do the birds leave at all, or go so far? Why do they face the privations, risks, and exertion of the journey? What drives their rapid fattening up without which they could not have enough fuel to reach their distant destination? Only raging appetite would fuel the fattening. Only an unquenchable drive to fly would make them *go* and keep going. The motivations and the behaviors presumably evolved because the Arctic summer provides more food than farther south, and so many species became adapted to be at home in that habitat. On the other hand, the Arctic provides little sustenance for most in the winter. The great migrations were shaped, then, by these imperatives.

I may be anthropomorphizing to suggest the godwits have a "love" of home, but although we can never know *what* they feel, it is hard to deny that they *do* feel. We can say that, along with the aforementioned cranes Millie and Roy, it is highly unlikely that conscious *logic* could drive them from one continent to the next. Animal behavior is first of all driven by emotion, although in us the emotion can be secondarily buttressed and/or amplified by logic. That said, we admire emotions that help us accomplish great things. We admire the drive and commitment that the birds show because our individual extraordinary feats pale in comparison to those of a godwit. The first lizards that sprouted feathers on their forelimbs could shield themselves from the rain and cold and may have been able to glide several meters, but for that they probably did not need drive related to homing. To fly nonstop for eleven thousand kilometers over open ocean, though, without taking a bite of food, a swallow of water, or a minute of sleep, is a mind-boggling demonstration of the epic importance of home, and of the ability and drive to return to it of even tiny birds.

Consider the example of a common European garden warbler, *Sylvia borin*. It is born in May somewhere on the northern European continent. It never in its life receives any instruction on when and where to fly. But two to three months after its birth it begins its flight in the night to Africa, where it has never been before. After reaching the Middle East, having flown in a generally southeastern direction, it shifts into a direct southerly direction and crosses the Sahara Desert. It eventually ends up somewhere in a patch of thorn scrub in perhaps Kenya or Tanzania, where it remains until spring. It then returns not just to the north, but perhaps to the same hedge in Russia or Germany from where it came, and after nesting there it again flies south to Africa to the same patch of thorn scrub where it wintered before.

Songbirds in North America do much the same. The Bicknell thrush, *Catharus bicknelli*, lives in the summer in the spruce forests on mountains not only directly adjacent to my home, but throughout the mountains of New England, the Catskills, and eastern Canada. It spends winters in the cloud and rainforests of Jamaica, Cuba, Dominican Republic, Haiti, and Puerto Rico. Christopher Rimmer and Kent McFarland and colleagues have been tracking these endangered birds in both habitats, to determine their home requirements. McFarland is the associate director of the Vermont

Center for Ecostudies and has banded nestling Bicknell thrushes in Vermont. The birds return annually to their same homes, and his first encounter of an overwintering thrush in the Dominican Republic turned out to be one he had banded nineteen months earlier in Vermont. He told me that capturing the same bird seemed like "winning the lottery while at the same time being struck by lightning. But for us naturalist types, much more exciting." On this occasion he broke out the celebratory Dominican rum on the very first night of that trip rather than toward the end of the fieldwork, as is more typical.

Routes of long-distance homing are now well known, but the *how* of the travel and the orientations to specific points of destination are still tantalizingly far on the horizon. The how is the most challenging of all to comprehend fully, because it literally involves everything about the animal at once — senses, metabolism, emotions, mechanics — all the physiology that runs the brain and the rest of the body. Solving such problems requires access and repeatability; animals don't migrate in the lab at one's convenience. Only one piece, or a few interrelated pieces, of the puzzle can be profitably examined at a time. Usually one animal species, by some quirk of its biology, provides access to a specific piece of information and another provides an opportunity for access to another.

The common rock dove or "pigeon," *Columba livia,* with its long association with humans, has provided clues to many aspects of homing. The same or similar general homing mechanisms of this "homing pigeon" could presumably also be used by migrant birds, and nonmigrating but far-ranging sea birds and turtles. Pigeons were well known since the Assyrians and Genghis Khan, who used them in war. Julius Caesar used them to send messages home from Gaul. They were used in the two World Wars and in the Korean War. Because of their attachment to home, they were ideally suited, as were swallows, for carrying messages, especially in wartime, as they were difficult to intercept and were probably more reliable for transmitting secret messages than the telephone and Internet are today. Fifteen-hundred-kilometer flights for birds in the U.S. Signal Corps are considered routine, and flights of twice that distance are recorded. One could release pigeons at any location and at any time and be assured they would try to return home, provided they were not too young.

One of the common sights wherever pigeons are kept is groups of them circling near their lofts in apparently aimless flight. Pigeons engaging in these flights are said to be "ranging" — they may be out of sight of the home area for a half-hour to an hour and a half. As in honeybees starting their foraging career, these flights are especially important for the young birds because during them they familiarize themselves with their home area.

Are the pigeons, like bees, using landmarks for homing? To test for this possibility, Klaus Schmidt-Koenig and Charles Walcott, both renowned bird orientation experts, put frosted contact lenses on pigeons' eyes to prevent them from seeing landmarks. To everyone's surprise, some of the pigeons, after being displaced, still managed to return to their home lofts. They flew in at high elevation and then fluttered down close to their home. The birds had apparently gathered some clues other than landmarks visible to us.

Through time and experience, and longer and longer ranging excursions, pigeons enlarge the area where they are at home. A working hypothesis is that "lazy" fliers, those that make only short flights, are unlikely to be able to home from long distances. Pigeon racers, who compete in the homing ability of their birds, bank on the knowledge that the longer the ranging flights, the swifter and the more accurate the homing ability. After about two weeks of ranging, the pigeon racer usually takes his or her pigeons farther away for each "training toss." Typically, the first training tosses are about thirty kilometers from the home loft. After three weeks the distance is increased to sixty kilometers, and then after another week to ninety kilometers. The birds' capacity gradually to increase their homing ability reinforces the notion that they are learning something about their home area, perhaps something like a "map" using some kind of landmark. Precisely what the birds are sensing at any one time that allows them to orient correctly to return home is not known, in part because it probably varies depending on the place and the situation. Although it is still not clear exactly how pigeons are able to home, we know that several senses are involved.

We have seen that some migrant birds stay together in family groups (geese, swans, and cranes), and that the migratory directions are learned from the parents, which expose the young to the relevant cues much like pigeon fanciers expose their charges with "training tosses" far from the home loft. The phenomenon of parental leading has been documented in whooping cranes, Canada geese, and ibises and extended by humans leading young tame birds to become imprinted on ultralight aircraft, in order to establish new migration routes. In most migrant birds, though, the migratory directions are inscribed in a genetically fixed "program." In either case, the migrants travel between one fixed territory in their summer home, and another in their winter home. However, presumably other, especially complicated mechanisms of homing are required in sea birds, which range far over the oceans and sometimes return to only a tiny speck, their natal island, after having wandered from it five or six years before. Do they build a map in their brain of some features of the ocean terrain that we can't see? In other words, do they see the ocean not as a flat, uniform expanse as we do, but instead as a featured pattern as of hills, valleys, ridges, and mountains in perhaps magnetic anomalies that inform them where they are at all times?

The one thing we now know for sure is that, like us and like bees, birds use the sun as a compass for homing. Gustav Kramer, a German ornithologist, perhaps the principal pioneer in homing behavior in birds, in the late 1940s tested the "sun compass" of pigeons in circular cages with food cups placed regularly around the periphery. The birds were trained to expect food in specific cups (directions). After the pigeons were trained, rotating the cage did not alter the direction where they sought food — except when the sky was overcast and the sun not visible, when they searched randomly for food at the different cups. Kramer repeated similar experiments with a well-known bird, the northern European starling, *Sturnus vulgaris.*

European starlings in Europe migrate south in the fall (though many or most of those now in Vermont and Maine do not), at which time they, as well as other migrants, enter a state of restlessness. Kramer coined the word *Zugunruhe,* meaning literally "migratory restlessness," to describe it. He first noted this behavior in his caged starlings, which were agitated and hopping around in their cages in the spring and tended to orient northeast. They oriented in the correct migratory direction when the sun was out, but as soon as the sky was clouded they no longer oriented in any one direction. Suspecting that, like the pigeons, they might use the sun to orient by, he tested his hypothesis by showing them the sun in a mirror and found that they then reoriented to the *reflected* sun. But the sun moves through an arc from east to west throughout the day, so how can the birds keep a constant migratory direction? Was the starlings' behavior a laboratory artifact?

In order to find out if starlings indeed adjust the angle of flight to the sun throughout the day, Kramer put his migratory restless birds into a room where they did *not* have access to sunlight. Instead, he provided a stationary light bulb to stand in for the sun. As predicted, if they used the light bulb as a substitute for the sun and possessed a time-compensated sun compass, the birds oriented increasingly to the left throughout the day. That is, they changed their intended flight direction with respect to the constant light bulb direction, treating it as though it were moving on the same schedule, of fifteen degrees per hour, as the sun does, and so they almost always faced in the "wrong" migratory direction in reference to the ground.

Kramer later lost his life while climbing a cliff trying to get baby pigeons to raise them for further experiments on homing orientation. But one of his students, Klaus Hoffmann, carried on his work. Hoffmann, who later worked at the Max Planck Institute for Behavioral Physiology in Germany, nailed the "time-compensated sun compass hypothesis" with another experiment in which he "tricked" starlings to misread the time from the sun's actual position. Given that the sun changes position fifteen degrees per hour, to keep flying in a straight line using the sun as a landmark, the bird has to know what time it is in order to compensate for the sun's shifting position. Hoffmann kept starlings in an artificially lit cage with a normal twelve-hour period of daylight, but with the lights coming on six hours *earlier* than actual dawn in the real (outdoor) day. These birds adjusted their activities to the artificial light schedule they experienced and expected food at a specific time in one specific direction in a circular cage, and their feeding time was of course six hours ahead of real or solar time. When his "clock-shifted" starlings were trained to expect food at their food cup in a specific direction and tested under a stationary light, they oriented ninety degrees (or fifteen degrees for every one-hour time shift) in a clockwise direction from their training dish. This experiment confirmed, by a different experimental protocol from Kramer's, the astounding hypothesis that the birds not only use the sun as a directional compass but, like bees, also consult an internal clock to correctly compensate for its rate of movement through the sky. Clock-shifted monarch butterflies also orient in the "wrong" but predicted direction, showing that they also use the sun as a "landmark" in migration.

This sophisticated behavior of insects and birds, however, does not explain the majority of homing orientation. Most songbirds migrate mostly at night, when they could not have access to the sun's location as a convenient directional beacon. (It is likely that small songbirds have to migrate at night because they need the daytime to replenish their energy supplies by feeding, whereas large birds, like huge airliners, have a longer flight range and burn much less fuel in relation to their body weight.) For a long time it was not known how, with neither landmarks nor sun available, the night migrants might orient. Yet orient they did, as experiments on warblers (Sylviidae) by Franz and Eleanor Sauer proved in the late 1950s.

Franz Sauer was a wide-ranging ornithologist from the University of Freiburg who had ties to many universities in the United States, including an appointment at the University of Florida in Gainesville. Highly inventive, he once made a blind to resemble a termite mound that allowed him to get close to an ostrich nest in the Kalahari Desert.

He and his wife, Eleanor, started their homing experiments by keeping warblers during their *Zugunruhe* in circular cages with a glass bottom and taking turns lying on their backs at night directly under these cages. They discovered that on starry nights the birds pointed to and attempted to fly in their cages toward the presumed migratory direction. On the other hand, on cloudy nights, the warblers were less active and much less precise in the directions of their migratory route.

The Sauers wondered if their warblers might be orienting by the stars, although this seemed like a long-shot hypothesis that would be difficult to prove. (The moon can serve well for night-flying insects, which only need to fly in a relatively constant direction for a short time. On theoretical grounds, however, the moon would not likely be an appropriate directional cue for long-range migrations since it appears at a different location every night and in different phases.) The Sauers found a simple way to test the hypothesis using a homemade planetarium. They brought the caged birds into their planetarium and saw that in the spring the warblers pointed north under the planetarium sky and in the fall they pointed south. When the planetarium stars were turned off, the birds became disoriented. Furthermore, when the axis of the planetarium was turned 180 degrees, the birds reversed their normal direction. The proof of the Sauers' hypothesis was convincing, leading to the next question: *How* do the birds "read" the sky? There are thousands of visible stars and numerous star patterns to choose from, and the positions of the stars shift throughout the night as the constellations sweep across the sky in a great circle.

A decade after the Sauers' work, Stephen T. Emlen from Cornell University continued to work on the stellar orientation problem, but he used a finch, the North American indigo bunting, rather than European warblers. One of Emlen's first improvements on the Sauers' laborious and often arbitrary determination of migratory direction was a clever innovation, which came to be called the "Emlen funnel." It is a funnel-shaped cage with an ink pad as a perch centered at the narrow bottom, and with paper lining the walls, and a screened top. A bird in *Zugunruhe* in the tunnel hops repeatedly from the ink pad on the floor onto the side of the funnel, leaving its every inky footprint on the clean wall, and hence it leaves a quantitative record of the intensity and the direction of its intended migration. Using this device, much data could be generated while the experimenter was asleep in bed rather than taking notes while flat on his or her back under the cage holding the excited bird. Emlen largely duplicated the main findings of star orientation that the Sauers had discovered, but he went on to identify both the star patterns that might be involved in the birds' homing and how they come to be recognized.

In the Northern Hemisphere, the whole sky appears to rotate

around the North Star, Polaris. Anyone with the least familiarity with the night sky can easily recognize the Big Dipper constellation and from it locate the somewhat faint North Star, which stays at a constant location throughout the night (and day) as the Dipper rotates around it in a counterclockwise direction. Emlen wondered if Polaris (or the star patterns near it, since they stay constant) might serve as a directional guide to birds, as it has been for centuries to human navigators. To try to find out, he systematically blocked out portions of his planetarium's night sky and found that many constellations could be obliterated without disrupting his indigo buntings' migratory direction. However, in order to orient correctly, his birds needed to see the star patterns near Polaris. As predicted, if the stationary portion of the night sky is used as a directional reference, clock-shifted buntings should *not* take a different bearing, as day migrants using the sun compass do. Furthermore, shifting the time of the night sky in the planetarium also did not result in reorientation, either — the birds still oriented in the same direction. These results thus helped to prove that the birds used the (stationary) north stars as their homing beacon.

The stellar orientation by indigo buntings turned out to be even more amazing than it seemed at first, because when Emlen tested birds he had reared in the laboratory without their having experienced a moving night sky pattern, he discovered they were unable to orient. Something was missing from their behavioral repertoire, and Emlen found out that it was learning.

The young birds need access to the night sky to experience the movements of the constellations. Later, after the birds have learned the positions of some of the constellations, they can determine direction even using a stationary (planetarium) sky. That is, like us, after the birds recognize the star patterns of the night sky, and know that those near the North Star stay in place and the others move around it, they can use any of a number of nearby constellations to infer the correct direction. To test his idea, Emlen raised buntings subjected to a night sky made to rotate about the star Betelgeuse instead of Polaris. As predicted by his hypothesis, under a stationary planetarium sky when they were ready to migrate, they oriented to Betelgeuse instead of toward Polaris on their northward migration. But another group of birds that he brought into the physiological state of summer (by photoperiod manipulation) and tested at the same time under the same planetarium sky oriented southward. Knowledge of the specific star patterns as such is thus not inherited, but the attention to them, the capacity to learn from them and respond to them, is.

Neither the sun compass nor stellar orientation explains all bird homing results, as William Keeton, a former entomologist specializing in millipedes at Cornell University, discovered with homing pigeons when he returned to his childhood fascination with pigeons in the 1970s. In his perhaps most famous set of experiments (after the university gave him a loft with two thousand pigeons to study homing), he attached magnets to pigeons' backs and found that the birds' homing ability was adversely affected, but only under cloudy skies. In 1974 Charles Walcott and Robert P. Green, of the State University of New York at Stony Brook, followed up by creating magnetic fields directly around pigeons' heads by gluing a wire coil on their heads and running current through it from a backpack battery. Their results confirmed that magnetism has little effect when the birds have the sun available, but under cloudy skies they apparently rely on the magnetic orientation.

The magnetic compass orientation hypothesis was almost concurrently confirmed for a small European migrant, the European robin, *Erithacus rubecula*. Instead of altering the magnetic field that the birds carried, Wolfgang and Roswitha Wiltschko from the Max Planck Institute for Ornithology in Germany altered the magnetic field around cages holding birds. The migratory-ready robins responded to those changes appropriately as expected if they oriented to Earth's normal magnetic fields. Since then, hundreds of studies show magnetic orientation to be so common as to be almost universal in a great variety of animals, although magnetic orientation and the sun compass are only two of multiple cues used in multiple ways.

Not far from Emlen's Cornell University in Ithaca, New York, the team of Kenneth and Mary Able found in the late 1990s that birds, in this case specifically savannah sparrows, *Passerculus sandwichensis,* a typical night-migrating passerine, use magnetic, star, and polarized light cues, and possibly also the sun, for determining and maintaining homing direction. But their preferred orientation reference is the magnetic, and even young hand-reared birds that have never seen the sky orient south in the fall by using the magnetic lines of force. However, visual cues (from stars and polarized light) can then be calibrated to the magnetic compass. In this case, information from star patterns and polarized light overrides the magnetic compass orientation. The sparrows learn to orient secondarily to the pattern of polarized light that varies in relation to the sun's position, and that is especially prominent at sunset. This system now appears to apply generally for songbirds, as confirmed by more recent research by William W. Cochran and colleagues using free-flying radio-tagged gray-cheeked, *Catharus minimus,* and Swainson's, *C. ustulatus,* thrushes.

Thrushes in spring migratory condition were subjected to an experimentally altered eastward-turned magnetic field during twilight (they were caged at that time before their release), a period when they naturally orient in a westerly direction. On release after dark (now in the Earth's normal magnetic field), they flew westward — although normally without the prior treatment they would have flown north. As predicted from their twilight eastward-turned orientation, they continued flying in the same (wrong) direction the whole night. But on the next and subsequent evenings, the radio signals tracking them showed that they had changed direction, now flying in the "correct" northerly migratory direction. The birds apparently take their direction cues from the polarized light patterns in the sunset direction, to recalibrate their magnetic compass direction specifically just before takeoff during each twilight. Thus, even when they migrate through areas of the Earth with perhaps local magnetic *anomalies*, they are not thrown off course.

The physiology of how the magnetic sensing works is still a mystery. Possibly iron-containing minerals align like a compass and cause mechanical deflection that is sensed. Presumably the cells or something in the cells swivels like iron filings do in response to a magnet. The mechanical displacement might then cause cellular depolarization, much like what happens to mechanoreceptors at the base of hairs when they are bent, as happens during hearing. Such a receptor has been claimed to be located in the snouts of trout, and in the upper bill of pigeons, while others suspect the magnetic information comes from the ears. Still another model being developed to explain magnetic information is linked to the visual system, and this one potentially provides vastly more information.

Animals "seeing" a map of the magnetic landscape may be more literally true than just hypothetical. Evidence is mounting that some birds can sense magnetic information optically — that the optic nerve transmits changes in magnetic fields to the brain, and light affects the responses to magnetic fields. That is, it appears that birds may "see" magnetic lines of force. What this means in terms of images is still unknown and would be difficult for us to imagine, although the evidence indicates that the magnetic lines of force are not seen in the way we read a compass, namely, along the directional component. Animals may additionally see magnetic lines of force along the vertical component. In the north, for example, the lines of force of the Earth's magnetic field are strongly directed upward away from the pole, at the equator they are nearly horizontal, and in the south they point downward toward the pole. Therefore, birds' ability to see lines of force gives them the potential to determine latitude.

More recently, the Wiltschko team examined a species of migratory Australian silvereyes, *Zosterops lateralis*. At migration time, their caged birds oriented predominantly in the northerly migratory direction. As expected, the birds reversed their headings from north to south when the vertical component of the magnetic field around them was reversed from the ambient normal. However, light intensity and light wavelength had a huge effect also; if the birds in the lab were put into bright blue light, they oriented along the east-west compass direction, and under green light they reoriented toward west-northwest. This result is puzzling and shows that there is more going on than we understand, especially since the directions taken in response to light changes were not reversed by inverting the polarity of the magnetic field.

A series of other very recent studies are delving into how the magnetic compass of birds is related to light-sensitive pigments in their retinas, since some of these pigments respond to a bird's alignment in a magnetic field. Such light- and magnetic-sensitive pigments are found in retinal cells of some migrants, but not in nonmigrants, and the neurons connected to these pigment cells show high levels of activity when the birds are orienting by magnetic information. Furthermore, recent studies of European robins by Katrin Stapput and others of the Wiltschko team indicate that specifically the right eye may play a dominant role. The robins were equipped with goggles; those with a frosted goggle that blurred their vision in the left eye oriented correctly to magnetic information, but those with vision from the right eye experimentally blurred did not. Thus, the observations in aggregate raise the possibility that birds may literally see ghost images of the Earth's magnetic field superimposed on their perception of images of objects.

How animals orient is now being examined also from a neuro-biological perspective. Recently the team of Le-Qing Wu and David Dickman from the Baylor College of Medicine discovered fifty-three neurons in the pigeon's brain stem that respond to the strength and direction of the surrounding magnetic field. Although the sensory organ from which these cells receive the magnetic information remains unknown, the researchers suspect it is the inner ear. The upshot of this rapidly expanding field is that rodents and humans have specialized neurons, called "grid cells," that in the brain produce localized electrical activity that shows up in functional magnetic resonance imaging (fMRI). These grid cells appear in localized areas that coordinate behavior, and such "cognitive maps" inscribe the directions and speed of movement of the animal. Some animals apparently carry a neural representation that keeps track of where they have been and are, in part by how fast they are going and hence how far they have traveled. Indeed, this neural representation can, in humans, be duplicated in a virtual reality arena that mimics the actual spatial movements.

The mechanisms I've described so far assume that the animals are adapted to a static world, one with dependable, though complexly changing, cues. But although the angle to the sun or to the North Star pattern is relatively constant, the *homes* of the animals in relation to them are not. Recent work with blackcap warblers in Europe shows a dynamic picture of rapid evolutionary change. Blackcaps are common breeders throughout northern and central Europe from where they traditionally migrate directly south to the Mediterranean and to North Africa in the winter. They nest in low garden hedges where they are conspicuous singers, and people welcome them each spring after their long migrations back north.

In the 1950s and 1960s, increasing numbers of the blackcaps were found to spend their *winters* in Britain and Ireland, where they had never been seen before at that season. Since blackcaps do breed in the British Isles, it was assumed that the winter-present birds there were simply some summer breeders that had stayed rather than migrating, because people had provided sustenance via bird feeders. But then on December 14, 1961, due to a cat and coincidence, the story suddenly got complicated. A resident in Ireland picked up a blackcap that a cat had just dragged in and noted that this bird carried a metal band on its leg. The band identified the bird as originating in *Austria,* and the finder sent it to the Austrian ornithological society.

The cat's catch strengthened a suspicion of Peter Berthold, of the Max Planck Institute for Ornithology in Radolfzell in southern Germany, that the blackcaps wintering in Britain might actually come from breeding grounds in central Europe. He and colleagues then conducted a historic study. Berthold caught blackcaps wintering in Britain, transported them to the Max Planck research labs, and kept them there until the fall when they were ready to migrate. He then tested them in cages and found that these birds, when in *Zugunruhe,* oriented westward, in the direction of the British Isles. In contrast, the locally breeding German blackcaps, which migrate to the south, oriented in his cages in *that* direction. Berthold then found that the respective direction the blackcaps choose is innate; it is determined by genetic programming. He crossed blackcaps bred in Germany that migrated south with those that also bred in Germany but overwintered in Britain, and these caged hybrids oriented at migration time in an intermediate, southwesterly direction (if they had been released they would likely have perished, because they would have landed on the Atlantic Ocean off the coast of France).

Continental European blackcaps overwintering in Great Britain, which apparently started as a rare phenomenon in the early 1950s, are now common. Tens of thousands of blackcaps now migrate each fall from their breeding grounds in central Europe to new winter homes, thousands of gardens throughout Britain and Ireland. The few initial blackcaps, which due to a genetic mutation initially flew to Britain rather than south from central Europe, had by chance come into a new home where they prospered, and then they multiplied.

Berthold suggests that the advantage of the mutation that resulted in the birds' migrating to and from Britain from central Europe could relate to the fact that the new migration route is fifteen hundred kilometers shorter than that of the south-migrating blackcaps. Perhaps the birds arrive on their breeding grounds in central Europe from Britain earlier and more rested than the longer-range migrants wintering at their traditional sites in Africa, and so they could pick and defend the best breeding territories before their competitors, those from the southern migration, arrived. Alternately and additionally, perhaps the wintering grounds in Britain are now more advantageous because of climatic changes and more bird feeders. The larger picture is clear: evolution of a new behavior has occurred almost instantaneously.

On reflection, these kinds of changes in homing behavior must have occurred in the past and probably still happen routinely. The birds are using magnetic orientation, and Earth's magnetic fields flip on average once every half-million years. These magnetic reversals are revealed in dated iron-containing rock that, as it hardened, "froze" the magnetic particles in it into the place of the then-existing Earth's magnetic field when they were still free to move in the molten material. The now-hardened rock that has erupted from the Earth's interior in the gradual spreading of the mid-ocean ridges reveals the record of the magnetic reversals that have been occurring for at least the past 150 million years. A magnetic flip can occur in the span of a century, and havoc in orientation may result, because suddenly all the "rules" have changed. Similarly, only ten thousand years ago, central Europe was covered with glaciers. There would have been no blackcaps at home there. When the birds invaded the cleared lands after the glaciers left in order to nest there, they would have had to migrate out in the winter and back in the spring. Relatively precise migratory directions had to be inscribed into their genes to ensure precise homing. The outliers were the safety valve from what would otherwise have been a straitjacket of inflexibility. As the landscape changed, those that changed with it prospered.

But why do those blackcaps that nest in Britain and Ireland still migrate? Why don't they just stay? Perhaps eventually they will. Perhaps a small mutation can change a migratory direction, but it would take a much longer time to eliminate migration behavior entirely because that behavior involves many responses and is therefore probably more deeply encoded. Migration as a whole involves timing, the restlessness to motivate long-distance flight, feeding binges, and sometime extreme fat deposition. All of this must require a large genetic program that likely cannot be shed by a one- or two-gene mutation. One thing we can be sure of is that the group of blackcaps that now spends the winter in Great Britain came from a small founder population.

Long-distance feats like the shearwater's, and those of other oceanic birds like albatrosses, became well documented, although the homing mechanisms were long a mystery. But before delving into the one study that chips away at the mystery of albatross homing, let us look at the homing of turtles, a far more "primitive" group of animals. Turtles are far older in evolutionary terms than birds. They are far older still than the dinosaurs and their relatives that last roamed the Earth and the seas sixty million years ago. Turtles in their present form were around over three times longer than that — about two hundred million years ago.

The late Archie Carr, a zoologist at the University of Florida, was one of the first persons to delve into turtle homing ability, and since 1955 he devoted his life to studying the green turtle, *Chelonia mydas*. Carr became the world authority on these reptiles and left a large legacy. His extensive tagging studies of turtles revealed a population that fed along the coast of Brazil and nested on Ascension Island, a five-mile-in-diameter pinpoint of land in the vastness of the Atlantic Ocean that is more than twenty-two hundred kilometers from the green turtles' feeding grounds. Mating takes place at the nesting grounds, and every two or three years the turtles make the trip from their feeding pastures near Brazil to their breeding place on Ascension and then back again. Carr realized that "this population seems to have evolved the capacity to hold a true course across hundreds of miles of sea — the difficulties facing such a voyage would seem insurmountable if it were not so clear that the turtles are somehow surmounting them." He believed that the ability to make the long sea voyages without landmarks, the capacity for open-sea orientation, is "the ultimate puzzle" in the study of animal homing. Although he deduced that the turtles "must have some kind of compass sense" that guides them in the open seas, he didn't know what senses were involved either in maintaining a straight line of travel or in identification of place.

Carr realized that to test for that "compass sense" he had to track the turtles' paths in the open ocean, much as Menzel had recently done with honeybees, equipping them with transponders and tracking them by radar over the countryside. Carr initially tracked turtles by having the animals tow a helium-filled balloon that would be visible even when the submerged turtles were not. But this method required staying with the turtles throughout their whole journey, and it would not be humane to leave the balloons attached indefinitely. He then tested the feasibility of equipping turtles with a radio transmitter. In 1965 Carr announced, "It may soon be possible for turtles to bear a radio transmitter and a power source" and "each time a satellite passed within range of the towed transmitter a signal would be received; these signals, rebroadcast to a control station, would allow a precise plotting of the position of the turtle." In this way the question of whether or not the turtles' homing resulted from mere improbable chance or true navigation could be answered. As in research with other animals, lab tests are almost always needed to complement field tests.

David Ehrenfeld, a biology professor at Rutgers University, equipped young green turtles with glasses in which only one of the two lenses was made opaque. Turtles so outfitted were lost: they kept moving in tight little circles in the direction of the eye that could see. Yet since the animals cannot see their thousands-of-kilometers-distant target as an image in the open ocean, they may instead see directions. Sea turtles are, in fact, notoriously myopic, although they orient to light and dark. It now looks as though vision may play a role not only in turtles' close-in detecting of images such as food, but also in their detecting of magnetic fields.

Insights gleaned from birds could later be reapplied to the problem of sea turtles' orientation. As Archie Carr had posited decades earlier, turtles must depend on some kind of compass orientation, even though by itself this capacity could not explain homing. Homing requires not only a compass, but also a "map." However, the team of Kenneth and Catherine Lohmann from the University of North Carolina have in the past two decades provided further insights into the homing mysteries of a sea turtle, the loggerhead, *Caretta caretta*, a species that nests on beaches in eastern Florida, returning there after swimming about fourteen thousand kilometers.

After loggerhead young hatch and emerge from their nest in the sand at night, they dash directly for the beach, guided by moonlight reflected off the water. Then, guided by the pattern of the waves after reaching the water, they attain the currents of the Gulf Stream, and then the Atlantic Ocean gyre takes them in a giant circuit into the Sargasso Sea. It is a grand migration of nearly ten thousand kilometers, at the end of which the one in thousands that survives to adulthood returns about twenty years later to a place near its home beach.

One may wonder if there is any homing at all — whether the return of one in four thousand to the natal beach is just random chance. Arguing against random chance is the fact that no loggerheads are known to make it to any other beaches, and also that predation is horrific. Most eggs are dug up and eaten by predators. The five-centimeter-long hatchlings must run a gauntlet of vultures and other predators before they make it into the water, where fish predation begins and continues even until they attain large size, when sharks eat them. The turtles now also face human predation, nest raiding, chemical pollution, and disorientation by light pollution from human beach lights. The studies by the Lohmanns, summarized in over thirty technical publications, indicate that loggerhead turtle (as well as salmon and spiny lobster) magnetic sense is especially highly developed, but that they use some half-dozen or more other types of information in homing.

The ocean and the coastlines have, like the land, regionally unique magnetic characteristics. For example, on a large scale the magnetic lines of force have both specific direction and intensity; those at the poles have strong intensity and strong inclination (dipping down in the north, horizontal at the equator, and south angling up), while those at the equator are weak and have little inclination. The superimposed local magnetic anomalies, if sensed, could identify locations. However, turtle hatchlings that have never been in the sea, when placed into magnetic fields characteristic of particular locations in the North Atlantic, still swim in their tank in the appropriate migratory direction.

The Lohmanns speculate that the specific instructions of the turtles' route are sensed from the magnetic landscape, and that the turtles are genetically programmed to respond to these magnetic cues correctly. The magnetic location of their home, however, is imprinted (learned), and so they know when they are close to home, and then they may be guided by scent, waves, and other cues to the beach of their birth.

Despite our ever-finer understanding of the micro, we still can't explain all the macro. The satellite-tracking studies of green sea turtles, *Chelonia mydas,* by Paolo Luschi, Floriano Papi, and colleagues at the University of Pisa in Italy confirm these animals' abilities to return to pinpoint targets on a straight bearing, presumably over areas where the magnetic information is never a constant. The turtles compensate for wind and current directions in their bearings. Their position fixing can so far not be explained by known navigational mechanisms.

Finally, I now return to the albatross, the bird that began my wondering if and how they know where they are. They are, like sea turtles, long-lived wanderers with fixed home positions within a to-us featureless sea. Suzanne Akesson from Sweden and Henri Weimerkirsch from France have shown, through long-term satellite tracking, that young wandering albatrosses average a flying distance of 184,000 kilometers their first year, have an apparently genetically fixed dispersal direction, and, after reaching a particular ocean zone, may stay there for seven to ten years before returning to their home site to breed. At least in the waved albatross, *Phoebastria irrorata,* homing is unimpeded by magnetic manipulation. Albatrosses with magnets attached to their heads continued to fly back and forth, like the control animals, the thousand kilometers separating their breeding and feeding sites. So their homing is apparently little or not at all related to magnetic orientation! We have learned much, but it has left us with the mysterious, magical, and miraculous. And the ultimate and perhaps unanswerable question, of whether any animals are conscious of where they are and what they are doing, remains.

SMELLING THEIR
WAY HOME

I miss your fragrance, sometimes I miss it this much that
I can clearly smell you in the air.

— Qaisar Iqbal Janjua

THE EIGHTEENTH-CENTURY INSECT BIOLOGIST JEAN-HENRI
Fabre raised caterpillars of the great peacock moth, *Saturnia pyri,*
and placed a female immediately after she emerged in the morning
from her cocoon "while still damp" in a wire-gauze-covered bell jar.
He had no particular plans for her but "incarcerate[d] her from
mere habit, the habit of the observer always on the look-out for
what may happen." What happened at nine that evening was a great
stir in his household due to what his family at first assumed was
bats but was in fact "an invasion of moths" that had entered a study
window. They were all great peacock males.

The incarcerated female lived for eight days, but more suitors
(each of which was denied physical contact with her and captured)
came every night, and each had a life span of only two to three
days. In aggregate, 150 males came into the house. Fabre knew

that these moths were rare and wondered how the female achieved her fabulous allure. In a series of clever experiments that should be the envy of a Nobel laureate, he disproved three hypotheses of how the male moths might have been attracted and proved that the allure was a species-specific scent (or an "effluvia"), one that is totally imperceptible to us. He later experimented with another giant silk moth, the lesser peacock, *Attacus* (now *Saturnia*) *pavonia*, which emerges more than a month earlier in the season than the great peacock moth, the males of which are attracted to females only near noontime instead of at night. He summed it all up in an exciting narrative that any bright eighth-grader could understand and ended up wondering about the difference in behavior between the two moth species, saying, "Let him who can explain this strange contrast of habits." At that time, nobody could.

The attractant scents, which we now call "sex pheromones," of female moths can attract males of the corresponding species from a distance measured in kilometers and can be routinely demonstrated. A freshly emerged female *Callosamia promethea* moth that I tied to a twig had by late afternoon dozens of males flying around her. Sex pheromones can also be used as a tool to kill moths. The bolas spider, *Mastophora hutchinsoni*, synthesizes a chemical that mimics the female sex pheromone of the cutworm moth, *Lacinipolia renigera*, and broadcasts that chemical in the evening when the moth males fly, thus luring them in to catch and eat. After midnight the same spider switches over to dispense the pheromone of another moth species, *Tetanolita mynesalis*, that is active only late at night. The bolas spider doesn't build a web but instead dangles a short piece of silk that has a gob of sticky glue at the end from one leg. When a moth comes near, the spider swings its "bola" at the intended victim to entangle and eat it. (We also employ synthetically produced moth sex pheromones to capture moths, but the lazy way — in conjunction with sticky tape rather than relying on tool use and manual dexterity.)

For many animals, scent is the primary window to the world, the cue guiding them to a potential mate, and to food. In fact, odor is an excellent means of "homing in" on a currently emitting source of a volatile chemical. But we don't generally associate the sense of smell with orienting to a home. The scents of home tend to be various and ambiguous, though, and might mislead more than lead.

Scent concerns the detecting and use of a huge number of different chemicals in an almost infinite variety of combinations, like a vocabulary of signs and signals with a large variety of meaning. The smell of ethanethiol, for example, is to us at very low concentrations (according to the 2000 edition of *Guinness World Records*) the world's "smelliest" repellent chemical. Yet, to turkey vultures, ethanethiol is apparently hugely attractive: they use it to home in on their favorite food, rotting carcasses. However, they are sometimes misled by leaking propane lines to which we have added ethanethiol in order to detect this to-us otherwise odorless gas. Isoamyl acetate, on the other hand, which to us has a pleasant banana-like smell, is what honeybees release when they are agitated enough to sting. It excites hive mates to attack, because if one bee is induced to sting, the bees' home is likely being attacked, and a mass defense is mounted. In the case of the honeybees, that sting results in their sacrificing their lives. Flower scents of various kinds cause insects to search for food. In this case, learning is often involved, as insects associate specific scents with specific flowers.

Nowhere is signaling by scent more important than in coordinating the societies of social insects. In honeybees, for example, the *lack* of one specific scent causes the workers to build wax cells to hold queen larvae, another scent emitted by a queen in flight attracts drones to mate, and still another of her scents depresses the ovary development of her workers so that she ends up as sole egg layer of the hive.

Scents are like the words of a private vocabulary; the babble of them in all the millions of species is largely "inaudible" to us, and hence their roles are often mysterious. Perhaps the most famous role for scent in homing occurs with ants that can, when they walk rather than fly, lay down scent, their "trail pheromone," with which successful foragers can establish a "follow me" communal trail that recruits nest mates to the food, and that can also guide them back.

Sahara Desert ants not only use the sun and landmarks in homing, but after coming close to their home, they orient by the local aromas that they have learned. The authors of a recent study scented desert ant nest entrances with different aromas and allowed the colony members to learn them, and then they moved some of the individuals to a new location and provided sites with the same scents as their parent colony. The ants then searched for their nest entrances at the scented locations, but only if *both* of their antennae were intact. Apparently, as in vision, they experience a scent gradient, providing them the ability to home in on scent location.

Honeybees also use scent in homing. When a honeybee swarm flies (guided by its scouts) to a new home site in the "cloud" of about ten thousand, most of the members have not before been to the new home where they are going, never mind where its dime-size "door" or entrance might be. But scouts perch at the tree-hole (home) entrance, lift the tips of their abdomens high into the air, open glands that then release a come-hither scent (which to us smells lemony), and beat their wings to broadcast an odor plume that guides the others to them. There may not be much of a breeze to make a scent trail for animals to follow, but the bees that plant themselves at the new home entrance create one by fanning their wings, and they may also align themselves into chains, which creates a directional airflow. Bees follow the scent trail and are led home, and until the bees have thoroughly learned where home is and can navigate to it, there will still usually be some that mark the home entrance with that lemony scent.

The "tubenose" sea birds of the order Procellariiformes (petrels, shearwaters, and albatrosses) use scent in finding food, and petrels additionally use scent to home in on their nests. Petrels are small birds that have probably experienced high selective pressure to avoid nest predation by gulls, and they have reduced predation by going underground. The male digs a burrow up to two meters in length, and the one egg of the clutch is laid at the end of the tunnel. Apparently as an adaptation to avoid gull predation, the birds arrive and depart from their burrows only at night, and both scent and sound are used in homing. Similarly, some experiments have suggested that pigeons may also use scent to aid them in homing in on their loft.

A potential problem for using scent as a marker of home concerns the reliability of signaling. Home can be normally associated with innumerable chemicals that can change on a minute-by-minute or a seasonal basis. How can scent "label" home so that we can reliably return to it by that signal? Which chemical or combination of them creates the smell we perceive to be the relevant one? And how can it end up becoming the *label* by a strong association in the mind to that place?

Imprinting (rapid irreversible learning) to a specific scent is important to many mammals. Many newborn mammals imprint on the unique chemical signature of their mother, and vice versa. Arguably this is attraction to a place, perhaps even a very specific one, such as mother, or even a teat. This phenomenon is especially important in herd animals, where the young are *not* restricted to a specific home, and the parents and offspring need to find and recognize each other within hours after birth not by a home or nest where they are located, but as individuals among a crowd of others.

Scent also coordinates mother-young interactions during nursing in pigs, sheep, and rabbits, and humans as well. Human infants imprint on the scent of their mother's breasts. But this scent is not a specific pheromone; in one experiment mothers slathered their breasts with a chamomile concoction while nursing their seven-month-old babies, and fourteen months later when the infants were given a choice to drink out of a chamomile-scented bottle versus one without the scent, they always chose the scented bottle. Those infants who didn't have the early contact with chamomile scent showed disgust toward it.

In general the main problem with scent for orientation and homing is that, while it is reliable up close, it is much less reliable from a distance. Still, scent can induce yearning, maybe nostalgia or a longing to be near and hence an imprinting to place.

Some mammals, including mice and bears, whose powers of detecting certain odors exceed ours probably by a thousand times, potentially also return home with the aid of scent. People have deliberately displaced both mice and bears (if they were "problem bears" or perhaps "problem mice") from picnic areas and pantries, to try to discourage them from raiding our food. Eastern deer mice have been reported to return to the same stump where they were caught from at least two kilometers and would presumably return from that distance to a human home with a pantry stocked with Camembert cheese as well. Most animals can't, like some bees, create their own scent-dispensing air currents. They must wait for a wind, and it must be from the right direction.

Relative to most animals, we seem to have a poor competency for scents, yet that is not certain because we may respond to some scents we are not consciously aware of. Experiments of men smelling sweaty T-shirts suggest that they find those worn by ovulating women more attractive than those of women who are not. That is, the scent may invoke feelings of attraction as well as repulsion. Although most of our feelings of nostalgia that are associated with home depend on conscious memory of what we saw, not smelled, scent is often a powerfully emotional attractant, although it is one not consciously remembered before being encountered. I recall returning one time to and walking up York Hill, my retreat in Maine, and, while coming to the curve of the path through the sugar maples near the top of it, smelling a subtle scent that suddenly altered my mood and made me feel good. The leaves had just fallen from the maples and the birches, and the scent may have come from the senescing or decaying leaves or from the mushrooms growing there. I don't know what it was, but it reminded me of the woods in Germany where I had lived and been happy as a young child. But being prompted by scent to remember and be attracted to a place is not a unique experience. A vivid description of this was written by Arthur D. Hasler, from the University of Wisconsin in Madison, for whom the experience opened up a lifetime career that led to solving the until-then-enduring biological riddle of how salmon return to the stream of their birth, after years at sea. He writes:

We had driven across the sage country and high desert from Madison, Wisconsin, where I had recently joined the faculty of the Zoology Department, to my parental home in Provo, Utah. Philosophically, this is about as far away from salmon country as possible. As I hiked along a mountain trail in the Wasatch Range of the Rocky Mountains where I grew up, my reflections about the migratory behavior of salmon were soon interrupted by wonderful scents that I had not smelled since I was a boy. Climbing up toward the Alpine zone on the eastern slope of Mt. Timpanogos, I had approached a waterfall which was completely obstructed from view by a cliff; yet, when a cool breeze bearing the fragrance of mosses and columbine swept around the rocky abutment, the details of this waterfall and its setting on the face of the mountain suddenly leapt into my mind's eye. In fact, so impressive was this odor that it evoked a flood of memories of boyhood chums and deeds long since vanished from conscious memory. The association was so strong that I immediately applied it to the problem of salmon homing.

That salmon swim up streams from the ocean to spawn in fresh water was known for centuries. But it was not known that they were returning to where they were born, or that some of these spawning migrations may be two to four thousand kilometers long. Given such incredibly long journeys and the commitment to make them — which often cost the fish their lives — ending at just any river or stream is not a good bet. There are adaptive reasons for the fish to come back home, to the place where they were born. The main one is the fact that the fish's predecessors had found there a proven good place for their offspring to be born and grow. There was, and hence is likely to be, gravel for the female to scrape a "redd" or nest where the water current brings sufficient oxygen, where the temperature is right, where there is food for the hatchlings as they grow, where the risk of predation is not overwhelming, and last where there are no insurmountable obstructions leading to these required stream conditions.

As mentioned, scent nostalgia (or memory) can work as an orienting cue for finding home only if the directions of the wind and water are reliable, and the direction of water flow in a stream is reliable. Fresh water has a different chemical content from the salt water of the ocean, and Hasler formulated the hypothesis that the young salmon can return to their ancestral home in each stream only if each one is flavored by a particular bouquet of fragrances to which the young salmon become imprinted before emigrating to the ocean. Years later when as adults they return from the sea, they could use that same scent as a cue for identifying and then homing to their natal stream.

Hasler's homing epiphany was first published in 1951 with his PhD student Warren J. Wisby. Their publication would later be followed up by more research, and it translated into a stunning record of science that combined laboratory and field experiments, ranged from chemistry to ecology, and concerned the natural history biology of salmon all over the world.

The test for the Hasler-Wisby hypothesis was bold and brilliant. The idea was to substitute a synthetic scent, one *not* found in natural streams, and to imprint salmon on it while they were young, and then to see if that scent could be used to decoy the adults during their spawning run to swim up a stream "flavored" with that chemical. This was a long-range and logistically difficult test, and a risky one as well, because it seems likely that salmon don't normally orient to a single chemical, but to a mixture. How then would one imprint young salmon to test their responses years later? Wisby's doctoral thesis was, first of all, to find and test *a* chemical that would be neither naturally repellent nor naturally attractive to the salmon, and the chemical he came up with was morpholine (MOR).

When Hasler and Wisby published their hypothesis and the planned experiment to test it, the only salmon populations where it could be tested were three thousand kilometers distant on the Pacific coast, not in Wisconsin, where they were. Since the Pacific was out of reach to them, they hoped someone else might become interested in the experiment and perform it. Nothing was done until almost twenty years later when an Atlantic fish, an invasive species, the alewife, became a nuisance fish in Lake Michigan. The Department of Natural Resources in Michigan planted coho salmon, *Oncorhynchus kisutch,* to control them. The coho thrived on the alewives and then spawned in the lake tributaries. Suddenly Hasler and Wisby recognized a serendipitous opportunity. With salmon in their own "backyard," the hypothetical experiment could actually be done, and Hasler's just-arrived new student Allan T. Scholz and other associates took on the challenge.

The test involved raising one group of the coho salmon hatchlings in a hatchery with water flavored with MOR and another group at the same hatchery in water flavored instead with phenylethyl alcohol (PEA). If the imprinting hypothesis was correct, each group of fish should have been imprinted on the respective unique chemicals found in their water. Then, to test if the scents would function in homing, the researchers released the hatchery-raised young fish of both populations into Lake Michigan. After these (marked) fish had grown to adults and were ready to migrate from the lake to spawn in streams, one stream entering the lake was flavored with PEA and another with MOR. These two streams and seventeen other (not artificially scented) potential spawning sites were monitored. The million-dollar question was, Which streams would their migrating salmon choose? The answer: over 90 percent of the fish chose the "correct" stream, the one flavored with the chemical they had known eighteen months earlier in the hatchery where they were "spawned." The conclusion was unequivocal: the salmon had remembered. They had been guided to the scent they had experienced while developing. Hasler finally had the definitive proof, and it came just two years before his retirement, by which time he had guided fifty-two doctoral students.

Andrew Dittman and Thomas Quinn of the School of Aquatic and Fishery Science at the University of Washington have recently extended the salmon homing studies with Pacific salmon (*Oncorhynchus* spp.) in Alaska. In a large lake system in Bristol Bay, reproductively isolated populations of this salmon are genetically differentiated and adapted to specific local conditions. For example, those breeding in shallow streams are smaller in size and have less of a hump, while those spawning in places with a fine substratum

lay smaller eggs. In these populations, separated into specific locations in the same watershed, the olfactory memory of the adults
coming back to spawn reaches back to the time of hatching. Returning adults discriminate among increasingly similar scent
sources until finding their eventual home site. Their sophisticated
and precise homing ability suggests a high selective advantage not
only to return to a suitable spawning ground of their species, but
also to find the almost exact site where they were born. But "almost"
exact is better than totally so.

> **Variability.** Few things in nature are as impressive as the
> apparent perfection of the navigation mechanisms that allow
> animals to perform their miraculous homing feats. Yet, as the
> French philosopher Voltaire famously pointed out two and a
> half centuries ago, *"Le mieux est l'ennemi du bien,"* usually
> translated as "The best is the enemy of the good." Perfection
> has a cost; it is difficult and takes a long time and a lot of ef-
> fort to achieve, and it can lead to a liability — a specialization
> that limits options. In nature, including homing, there are
> instead mechanisms (such as sex) that create variability and
> act specifically to avoid present perfection in favor of com-
> promise for future continuity in the face of unanticipated yet
> inevitable change. Nature and Voltaire, in other words, are in
> agreement that ever more perfected mechanisms generally do
> not achieve the best long-term results.

"Scent" of the water as such gives no guarantee that any given
stream has suitable spawning areas, such as the gravel beds required by salmon. Some streams have neither suitable spawning
nor feeding areas, while others may have both but be blocked by
impassable waterfalls (and more or less guaranteed these days, also
dams). Only that stream from which the fish came is sure to have

access to both spawning and feeding grounds. The fish may make only one spawning migration in its life, and the homing journey may cover hundreds of kilometers, so if the wrong stream is taken, the fish may forfeit reproduction entirely. It has therefore little choice (if it could choose) but to return to the natal stream. But as I have mentioned, while Arthur D. Hasler and colleagues proved that most coho salmon are imprinted as young fish on the scent of their home stream, a small percentage of them chose another, or "wrong," stream.

In general those fish that choose the wrong stream do indeed have a lesser chance of reproducing than those that do not, and as a consequence, one might on the face of that predict that there should be selection for ever-better homing and home discrimination. However, a locking in to current perfection is almost by definition disastrous in the long run, as long as change is a reality. And it usually is, given enough time. Variety is nature's way of creating options in the face of future change, and in the long process of evolution, sex has evolved as a mechanism that scrambles the gene pool to create variety.

Over evolutionary history there were scenarios that eliminated the breeding opportunities at a salmon population's home stream (such as a volcanic eruption poisoning the water, a watershed drying out, or a landslide creating an impassable waterfall), and at the same time, due to climate change, other waterways were created. The whole population at a stream could be wiped out. If, however, a salmon that spawns a thousand eggs has several offspring that end up at the one "wrong" stream that then turns out to be "right," it could reap a huge reproductive bonanza at little cost because the new and uncontested resource would yield a whole population of descendants. That is, in analogy with the homily to not put all one's eggs into one basket, "imperfect" homing can provide life-saving opportunity for some of the offspring, which then become the standard-bearers of a new population. Indeed, I think I found the perfect example of this right next to my cabin, but with amphibians.

My cabin is on a big steep hill, and it is deep in the woods. It is far from anything resembling a wetland where you would expect to find frogs and salamanders. Yet, I've found a red eft, a spotted salamander, and green and leopard frogs there. They looked as if they were lost, because they were in the "wrong" place, far from where they could live in the lowlands, and uphill at that. Later, though, it looked as if there is a method to such wandering. I happened to dig a hole about five meters across, for a possible decorative "pond." It filled with water because there was a catchment of rock underneath. The first spring a couple of wood frogs found it and stayed to call. Now, the wood frog chorus from there is deafening every spring. During one later spring I heard a spring peeper there also — these frogs are no larger than my thumb, and I could not imagine how one of them had apparently traversed kilometers of forest to arrive at this spot. But several years later there was a small chorus of them. Then, the same thing happened with green frogs, and with salamanders. All of the amphibian species that come there now to reproduce also home to their places of birth, but not all of the individuals do so. Some disperse, and when they do, they risk much but gain a small chance to "win the lottery." Somehow a few individuals had found the right spot in a sea of unsuitable possibilities. But I doubt that either their willingness to wander, or their ability to find and recognize what they have found, is not associated with evolved behavior.

PICKING THE SPOT

In late April and early May, at the end of the Maine winter, the wood frogs are chorusing and mating and the birds are starting to rush north as though sucked up by a vacuum. The woods ring with their song, and the drone of bees fills the air. Willow, maple, and other tree flowers are opening, and bumblebees are fueling up with nectar, getting ready for the most important decision of their lives: where to build their nests. You see them here and there, zigzagging a few centimeters above the matted leaves and grass where there may not be a flower in sight. When there are flowers, these bees often ignore them. Now and then one of them lands and walks on the ground, then lifts off again, flying this way and that, to continue her close ground inspection at another place. Occasionally one of these recently overwintered bees crawls into a hole or crevice and disappears from view, only to emerge a minute or two later, wipe her antennae with her front pair of legs, and then fly on.

It is sometimes easy to follow a bumblebee for several hours, but only in the summer while she is foraging for nectar and by then also pollen as food for her larvae, making only short flights in an open field blanketed with flowers — because bees can work for an hour or more before gathering a full load of nectar or pollen. But these early spring bees show interest in the ground (and, in some species, in tree trunks), and they may fly for kilometers and keep it up day after day if they can tank up quickly at a patch of sugar-yielding flowers. They are all "queens," females who mated the previous fall. They are hunting for something, but it certainly cannot be mates. They have no need for more sperm for the rest of their lives. Instead, they are searching for a place to home, where they will spend the rest of their one year of life, to raise a family of perhaps hundreds of sons and daughters. Not just any place will do.

In northern climates, the house-site-hunting bumblebee queen must be fussy. A queen who chooses a wrong home site forfeits her entire reproductive life. She needs to find a shelter that will remain dry and insulated from the not-infrequent cold rains and frosts. I do not know how these bees evaluate potential home sites, but they choose small dark dry cavities with fluffy material in them; many species of bumblebees take over existing old homes such as mouse and squirrel nests. Sometimes they even evict the nest's rightful owners. One time I found an Arctic bumblebee queen, *Bombus polaris*, on Ellesmere Island taking over an active snow bunting nest of dry grass lined with feathers. These far northern bees have demanding requirements for a home, but honeybees have an arguably even more demanding task in finding a home site.

Honeybees originated in the tropics where they evolved to maintain a large year-round colony. They require a solid and roomy home, one that accommodates a population of thousands. A bird or mouse nest won't do, and the primary reason they are now able to live in the north at all is because of highly refined homing behaviors. Bumblebees hibernate in the winter, and the queens, the colony founders in spring, burrow into the ground; the colony itself does not survive beyond fall. The honeybees' home must serve as a shelter not only through summer, but throughout the coming winter or perhaps numerous winters as well. Their nest site must protect them from deep frosts, and it must be spacious to hold large honey and pollen stores in the winter. The colony needs these stores, not just to survive the winter but also to rear young then. The young not only are fed but must be kept warm in order to grow. This means the home must be heated, which requires a constant use of fuel. The colony starts rearing young while there is still snow on the ground to build up a large worker population to exploit the spring bloom. That is, honeybee homes must be large enough to accommodate not only tens of thousands of occupants but also to store huge amounts of honey to feed and fuel their heat generation through the winter. And because they contain the rich honey stores and brood nurseries that tempt predators, they must additionally be heavily defended. The upshot is, to ensure adequate protection from both the weather and predators, honeybees in the north temperate region must nest in a securely enclosed cavity, one where they can pack in perhaps twenty-five but up to fifty or one hundred kilograms of honey and pollen, and also have room for nurseries for rearing young, plus hold the population of several tens of thousands of bees.

In the wild, the usual site chosen by honeybees is a large hollow tree. Unlike bumblebees, where only one bee founds the colony, it takes a swarm of honeybees to found a colony in a new home. A swarm consists of about half to two-thirds of the home occupants and their old queen, which depart usually shortly before a new queen emerges (if the old queen does not leave, there is a life-and-death battle between her and her daughter). Perhaps a third of the bees in a colony stay after the swarm leaves, and they will soon be supplied with the newly emerged queen.

With honeybees, the decision of *which* home site to use is not made by the queen, as it is with bumblebees. It is instead a social decision that involves her sterile daughters, some ten thousand individuals, and that decision is based on an evaluation of alternative choices that representatives of the colony offer up. Each site is evaluated, and the results inform a democratic process and a unanimous choice.

The story of honeybees' home choice arguably began with Karl von Frisch, when he elucidated the honeybee dance language. But in the early 1950s, his student Martin Lindauer used knowledge of the stunning von Frisch discovery as the lever that eventually solved part of the mystery of how honeybees choose their home sites.

Lindauer saw bees doing their waggle dancing on swarm clusters — on the surface of the mass of bees that forms after the bees of a hive have left their overcrowded home and before they have found another. Successful foragers inside the hive routinely use such dances to indicate the direction and distance of food. The foragers are usually dusted with pollen, and they carry two pollen packets in the two corbiculae (hairs adapted to hold pollen) on their hind pair of legs. But the bees Lindauer watched on the swarm clusters carried no pollen; they were instead smudged with dust and soot. Since swarms are groups of bees in need of a new home, Lindauer deduced that these "dirty dancer" bees were "scouts" that had examined cavities in the ruins (this was in postwar Munich) and were now advertising potential new home sites.

Lindauer's subsequent work based on his acute observation was published in the now-classic paper titled (in translation) "Swarm Bees in Search of a Home." In it he revealed that there were often several potential home sites being advertised simultaneously by different bee scouts, yet all the scouts eventually agreed on just *one* site, the best of the lot. After all the dancers had reached unanimity (were all indicating the same site), the more than ten thousand bees of the swarm left en masse and flew directly to the nest site. Being able to read the scout bees' "language" that encoded the locations (approximate distance and direction) of these sites, Lindauer was able to predict the future home locations advertised by the scouts and was even able to arrive at the bees' chosen home site before they themselves arrived to move in! But this left a fascinating question: *How* had the bees reached a consensus, and how had they achieved the social coordination to get to the chosen home that most of them had never seen?

When I was teaching insect physiology and conducting research on temperature regulation by bumblebees and honeybees at the University of California at Berkeley, I was puzzled by how it was possible that all of the bees of a swarm took off together. I knew that their flight muscles had to be heated to about thirty-five degrees Celsius before they could fly, but most of the bees in swarm clusters that I examined were in near torpor, because when I shook a swarm off a branch to capture it, most of the bees dropped straight down onto the ground (or into a container that I held under them); they had apparently not yet shivered and warmed up. My measurements showed that only those bees that were in the *core* of a swarm were warm enough to fly, while the rest, which was most of the bees, were not. So how could *all* the bees of a swarm fly off at once at one specific time, as they needed to do in order not to be left behind when the swarm left to fly to its new home?

My wild swarms (procured mostly with the help of the Berkeley fire and police departments) were set up to hang at the windowsill of my upstairs lab over the Wellman Hall Entomology Department parking lot. I then installed temperature sensors in numerous locations in these swarms, recorded the temperatures electronically, and printed them out on a chart recorder so I could view them continuously from inside my lab. Days often went by with only a few bees coming and going, while the swarm, the mass of them, remained too cold to fly. But then suddenly one day the *outside* layers of bees on the swarm cluster started to warm up, and all these bees got hotter and hotter. And then, shortly before a swarm takeoff — the time during which a consensus of a single future home location is reached — the temperature of the swarm periphery finally reached the same as that of the swarm interior, which was also the body temperature the bees required in order to fly. The swarm then quite suddenly dissolved into a cloud, which enveloped much of the parking lot below my window, and not very quickly coalesced, and then took off in one direction, presumably to their newly chosen home. This proved what I wanted to find out. For various reasons my swarm studies were soon terminated, and I went on to other things. But, as almost always, answering one question raises others.

One of the first questions other scientists asked was: How do the bee scouts evaluate the suitability of a potential home site? How do they decide if one potential home site is better or worse than another? It took another "bee person," Thomas D. Seeley, to take this on as a PhD project in 1975 at Harvard University. Seeley wanted to find out, as he put it, "what makes a dream home" for bees. Almost four decades and two important as well as entertaining books on bees later, he is now at Cornell University and still making important discoveries, after having plumbed the bees' collective intelligence in their life-and-death decision concerning the best possible home available within a ten-kilometer radius from their original home.

Seeley broke the home-hunting process down into six basic steps: scouting, reporting, advertising, debating, rousing, and piloting to the agreed-upon home site. He imported swarms to Appledore Island off the coasts of Maine and New Hampshire. On this small deforested island, the bees had no other choice of home sites except the boxes of various dimensions and qualities that he provided them. Not only could he watch what happened at the swarm clusters, as Lindauer had done, but he could here also concentrate his attention on the potential home sites themselves. The scouts, which were marked with tags so he could differentiate individuals, did not dance for some potential home sites they had visited (observers were stationed at the potential home sites to be able to report which individual bees had visited) but did so vigorously for others, and so he could then determine what qualities the bees valued.

Seeley determined that the scouts "paced" out the insides of his boxes, the potential home sites, apparently to assess volume. Home volume, he found out, was only one of several relevant criteria for bees in their home evaluation. The new home must of course be large enough to accommodate all the bees, as well as to provide eventual storage space, as we've learned, for at least twenty-five kilograms of honey (mostly for winter heating fuel) and pollen (baby food), and space for rearing young once the queen gets going and lays some thousand to a thousand and a half eggs per day. On the other hand, the home cannot be too large or the entrance too big, or heating it in the winter (by shivering and huddling to retain as much heat as possible) becomes impractical. If the cavity conformed approximately to a usable home site, the scout who found it returned to the swarm cluster and danced on its surface to indicate its quality (by her dance enthusiasm) and its location.

Home site evaluation, though, was just the beginning. The next problem Seeley tackled was how the bees reach a consensus, since the various scouts danced to indicate different home sites they had discovered. The consensus making, he discovered, starts with dancers who have found one potential home site also attending the waggle dances of other scouts advertising other (different) potential home sites. A process of elimination ensues as scouts who found superior home sites dance longer and more vigorously and thus recruit more converts. More and more scouts become converted to a popular new home site, until a "quorum" is reached. That is, the bee is a realist. She is not wedded to her own ideas simply because they are hers.

At that point, the scouts, upon returning to the swarm, stop dancing and instead make "piping" sounds/vibrations (by a shiv-ering-like contraction of their flight muscles), which I presume is a symbolic preflight warm-up behavior. In any case, it is the signal that encourages the other bees to shiver and warm up also, and so the swarm becomes ready to fly to a new home.

Seeley discovered that after the swarm is warmed up and flight ready, the scouts give a liftoff signal. They make mad dashes through the mass of clustered bees in "buzz runs" that signal it's time for takeoff. The buzz runs look like a symbolic takeoff, and I suspect the piping also comes from flight muscle vibrations arising from preflight warm-up in which the wings are partly engaged with the muscles, as it were, with the clutch no longer all the way in. During this activity, the scouts nudge and bump into other bees, inducing them to take flight. As a result, the swarm dissolves into a diffuse milling cloud of bees, as I saw often over the entire parking lot at the University of California from my upstairs window, and often in some surprise (if not also a tinge of alarm).

After the swarm is airborne, the former scouts, all of whom know where the chosen home site is, "streak" in rapid short flights through the milling crowd. The direction of these streaker flights is toward the previously chosen home site, and their flights induce the other bees to fly in that same direction. But the bees need still one more signal to proceed; they need to be sure the queen is with them.

The streakers (former scouts) have been many times at the in-tended home, whereas the queen and the rest of the bees have never been there. The queen plays no apparent role in guiding the swarm, except that her scent is required to assure her presence for others to continue their journey to their chosen home.

It was long thought that the streaker bees could be guiding the swarm by their "come here" scent from the Nasonov glands at the tip of the abdomen. But a study by Madeleine Beekman and colleagues, in which these glands were sealed shut in all the scouts, indicates that the bees are directed to the new home site without the need for this scent.

But as already mentioned, scent *is* the "sign" *at* the house entrance. Except for the few scouts, none of the thousands of bees can have any notion of where the entrance — a small hole in a large hollow tree — is. On their own, each of the thousands of bees would be unlikely to find it. Nevertheless, it is impressive to see how they quickly enter; indeed, they literally stream into their new home. That is possible because the scouts alight at or in the home entrance, raise their abdomens into the air, and then release the scent from their Nasonov glands. They beat their wings, which creates an air current that spreads the scent, and the swarm follows the scent plumes to its source, the entrance of their new home.

Honeybee real estate like any other depends on supply and demand. When honeybees were first brought to the East Coast of America by the colonists, they had at first many hollow trees to use as home sites. But eventually they had to settle where others had not. They had to occupy hollow trees that had not been already occupied. At first, picking a spot was easy, but the demand created an outward pressure, and America provided an "experiment" on a continent-wide scale showing the effects of only a modest move by offspring from their parental home. There were no honeybees in North America before the colonists came. One jumping-off point for the bees' expansion was in Virginia in 1622, and another occurred in Massachusetts in 1640. The bees, without any bias to move away from home, spread westward. Their rate of westward expansion exceeded that of the human colonists; wherever the settlers went, the bees, which the Native Americans called "the white man's flies," had preceded them.

By 1770 honeybees had spread to the banks of the Mississippi River. Everywhere the colonists went they found "bee trees," their source of wax for candles, ambrosial sweets, and the means to make intoxicating mead. Sugar was, as it is now, a highly valued commodity, and, as during the previous thousands of years, it was still available only from the nectar of flowers, and only through the labor of bees. Thus, the settlers' home-making in their westward expansion was intimately associated with, if not facilitated by, the bees' home-making.

There are thousands of bee species, but since most are tropical and relatively few are social, the honeybee way of evaluating a potential home site is likely unique in scope and complexity. Nevertheless, the specific requirements for bees' home sites vary enormously. Many tropical social honeybees, for example, build their nests in the open on cliffs, as they have less need for shelter than northern bees. Solitary bees, depending on the species, may choose a sand-bank and dig a tunnel for a subterranean home. Others use a narrow tube such as a reed stem or an insect burrow in wood. One, the mason bee, makes a little home of mortar. Each type of nest site is as though inscribed on the animal's nervous system as a "template," and the bee then matches any number of specific features of that template in finding a home site and making a nest. As with homing, the criteria for home site selection are tailored by evolution to the animal's biology in its unique environment.

Choosing a home site is an almost universal problem animals confront, and there are two main possibilities. For birds, one is to search far and wide for the best possible habitat in which to set up a home. The sampling must often be relatively brief because the bird's life span is short, and potential home sites are infinite. As in insects in a seasonal environment, the right spot must be found quickly, and most birds have an innate "template" of what the habitat looks like that is suited to their behavioral program and their needs. They can also bypass the searching and testing of specific habitat type and rely on the experience of others of their kind, taking their cues from them. If others have found an area good because they live there, it "must" be good real estate, according to this "logic." Colonial sea birds are an obvious example — if two islands look similar but one has a lot of your species nesting on it, it is likely to be the one without foxes, and the one containing other potential life necessities near there as well. Planting artificial models of Atlantic puffins, *Fratercula arctica,* on islands along the Maine coast has been successful in attracting these birds to breed where they had been absent prior to seeing others there. To test if songbirds are also attracted to habitats where others are present has involved broadcasting the species' song in an area, and then seeing if others of that species settle there. Some songbirds prior to migration in the fall scout for sites to come back to in the spring and are attracted to the dawn chorus of others of their kind. In that way they have a specific goal and are ready to start nesting soon after they arrive back, so broadcasting after the breeding season plays a role in recruitment for settlement. However, in a study of the black-throated blue warbler, *Dendroica caerulescens,* broadcasting songs of the species during the part of the season when the birds return to settle into their home territories did not have an effect on recruitment.

In some ways picking a home site is simpler for birds and us than it might be for honeybees, because we act more as pairs or as individuals and so need no elaborate procedures for consensus making for this decision (although I am sure this can be debated). Nevertheless, we still have the necessity of evaluating different possibilities and making a choice. In birds, it is usually the female who is the home site choice maker. She as the egg layer is by default the ultimate decision maker. She prepares the egg-laying place, which may be no more than a scrape on the ground. Almost any site will do, provided it is hidden from predators and offers sufficient support on which to build the nest. A robin's nest on a branch, for example, will not stay up unless it is built on a sufficient foundation. A robin may repeatedly carry grass and strips of bark onto a slanting limb, from where these items keep slipping off and falling to the ground. I once watched one do this for a while, and she eventually stopped, to try again elsewhere. She was trying to satisfy an urge, which in this case was to place one item on top of the other. If many items had been piled up, she would have sat on them and turned, placing more of them at the edges, as gradually a nest would have been formed. One result sets the stage for the next step. But the first step is always choosing the right spot. The more propitious the original choice of home site, the less time is wasted, and most birds are very good at picking just the right spot. They may go through a process of "seeing" many potential home sites, and then narrowing them down until they come to one that seems "just right." Of over a half-dozen broad-winged hawk nests I have seen, all were over ten meters in a triple fork of a hardwood tree. Peregrine falcons' nests are on cliffs, if available. Dozens of winter wrens' nests I've seen in New England were all in the roots of windblown tree tip-ups, every vireo nest that I have ever seen was in a fork of a horizontal twig, and those of brown creepers are always under dead peeling bark.

Homes may be built for safety from the elements and from enemies, especially with respect to the most vulnerable, the young. Consequently, safety is always of critical importance for most animals' choice of a home site, because eggs and young are easy pickings for almost any predator. How the sites are evaluated depends on a combination of innate preferences, but learning is also involved. The Canada geese that I have observed now for over twenty years in a series of beaver bogs in Vermont show a clear pattern that may be representative of some species. The first time the geese started to nest in the beaver bog by our house they chose a sedge hummock in a large open area of many such hummocks, and they were successful in bringing up goslings. Now, however, there are many geese that each year come to try to nest in the bog, and predators, perhaps coyotes, raccoons, mink, and otters, have learned about the eggs that can be found there. Now every goose that builds its nest on one of these hummocks ends up with its clutch destroyed. There is only one spot where the nest is never (so far) destroyed, and that is on the top of a beaver lodge that is surrounded by water. The geese fight viciously to occupy that safe spot. But it is not islands as such that they recognize as safe. In some other areas Canada geese nest on abandoned raptor nests, on cliffs (as along the Mississippi River), on platforms put onto trees, and I have a report of them nesting, in Cambridge, Ontario, on the third story of an industrial building, and in large flower pots alongside pedestrian paths. When a pair in that area choose to nest on another site, their nest is commonly raided before the eggs hatch.

Safety of the nest site can also involve neighbors. In one study by Harold Greeney and his assistants in the Chiricahua Mountains of Arizona it was found that the nests of black-chinned hummingbirds, *Archilochus alexandri*, were clumped in the immediate vicinity of accipiter hawk nests. These bird hawks do not hunt hummingbirds, but they do hunt jays and squirrels that prey on hummingbird nests and create a predator-free habitat. The hummingbirds would not need to consciously or deliberately seek hawks to nest near them. They merely need to return to nest where they were successful and move on if they failed. Similarly, in the tropics, some species of birds routinely build their nests next to the homes of aggressive stinging wasps. The wasps attack anything that, like a monkey, would jiggle the branch from which the nest hangs. In a variation of this theme, some small birds in temperate regions make their home directly in the understory of the huge stick nests of eagles and other large raptors that would probably not have access to their "subletters" because they live on the bottom floor, and besides are just "small fry" and not on their hunting menu.

Rules change when the nest site is no longer arbitrary and becomes a unique and valuable commodity. The nest can be a valuable item manufactured at great cost and skill, in which case it can become a nuptial offering of competing males. Thus, in weaverbirds and woodpeckers, the males do most of the site preparation — they build at least a partial nest or hammer a cavity out of solid wood to nest in, and the female inspects it and in choosing it, chooses him.

PART II

Home-making and Maintaining

The homing mechanisms of other animals are often mysterious to us because they are usually not apparent except through experimentation, and they involve sensory capabilities and neural processing that we lack. But one of the most widespread, diverse, and sometimes spectacular aspects of animal behavior is home-making, and the process is not just inferred. The results of home-making behavior are preserved as discrete artifacts, tangible records of various steps of behavior. Curiously, animal behavior texts seldom mention home-making, even though animal homes and the behaviors contributing to the theme are essential features for survival and reproduction.

ARCHITECTURES OF HOME

Home-making is practiced by animals regardless of their position on the evolutionary tree. The home may be a structure that shields both from the elements and from a diversity of enemies, and it may be especially important for rearing offspring until they are ready to face a challenging environment. It is most prominently displayed in insects and birds but is also found in various forms in mammals, spiders, crustaceans, fish, and some reptiles. Home-making is usually species specific and in many cases seemingly idiosyncratic, perhaps because of its various simultaneous and often conflicting functions. It does not invite making other than obvious generalities except for the main one: whether mammal, bird, or insect, home-making functions for security mainly for raising young, and in some cases this has been the crucial step toward an overlapping of generations leading to a truly social lifestyle.

Whether the home is a cavity with solid walls as with honeybees, or consists of underground tunnels and cavities as with naked mole rats or of "castles of clay" as with many termites, making the home may require excavating; piling up sticks, rocks, fibers, and a nearly endless variety of other materials; and assembling and cementing them together with silk, mud, feces, or saliva into any of a huge variety of structures that are often stunningly complex and beautiful. If they were not for specific uses, some of these creations would be considered art of the most exquisite sort. Nevertheless, some are indeed "for show" also.

Home-making may involve using existing shelters. But it can also involve making the very material that is used to build with, such as a bird that makes its nest entirely from its own spittle, frogs that make a nest from foam created with their cloacae, cicadas that make homes from abdominal secretions whipped into a foam by a kind of breathing-apparatus-cum-air-pump, or caterpillars that make a communal nest from their silk, to trap solar heat on cold mornings.

As elegantly described and illustrated in Karl von Frisch's book *Animal Architecture,* home-making is most "advanced" (changed from its ancestral form) in some of the bees, wasps, ants, and termites, the latter recycling their own feces as binder to construct their homes which may house millions of occupants. Termites' homes may in some cases be up to seven meters tall and are reminiscent of cities in miniature. They are constantly repaired and may last decades. Homes such as those of the African *Macrotermes bellicosus* attain air conditioning by a system where hot air, which rises, sucks in cool air from the bottom. Millions of inhabitants live in a temperature-controlled environment created by this thermosiphon temperature control system, without ever venturing outside. The compass termites, *Amitermes meridionalis,* of the scorching treeless Australian steppes achieve an efficient kind of temperature control by building homes up to five meters tall and aligned so that their broad flat surfaces receive the rays of the low morning and evening sun, whereas the tops of the mounds are sharp and pointed and present the least reflective surface area to the searing midday sun. In areas where heat is a problem but flooding is not, ants, termites, and many rodents build subterranean homes where active cooling can be dispensed with. The huge communal homes may contain (as in some ants) gardens that grow food crops and "cows" in the form of aphids that are "milked" for food secretions. Ants' homes, depending on the species, may be no more than a hollowed-out acorn, leaves silked together to create a cavity, or in the case of leafcutter ants, *Atta*, nests so huge that they would dwarf a person who might enter one. These underground spaces contain fungus gardens, where the leaves the ants harvest to grow fungus that they feed on are kept, and also contain spaces for refuse heaps, nurseries, and a chamber for the one female in millions that is allowed to be sexual. The variations are endless, and the mechanisms of how they evolved from just seeking any available shelter to making "castles of clay" are often inscrutable.

Let's continue the story of honeybees by looking at their home-making. As soon as the colony moves into a tree cavity that will become its new home, the bees begin to work their mandibles along the walls, loosening any debris and rot, dragging it out, and flying off and discarding it. When solid wood remains, they smooth it by coating it with resins (collected from tree buds) mixed with wax. Small holes that might leak heat are plugged with the resin as well. At the same time, some bees begin to forage and return with their honey stomachs filled with nectar. A critical need now in the home-making is for wax to build the honeycombs.

For over two thousand years, ever since Aristotle wrote on the subject, it was believed that honeybees gathered the wax to build their home from flowers, where they get their nectar and pollen. But after a long history we learned instead that young worker bees synthesize the wax, peaking in their production of it at around two weeks of age. The wax is made from sugar by glands between the abdominal segments. Foragers coming back to the nest without a place to deposit nectar indirectly sense the need for storage containers for honey and are stimulated to convert the sugar they carry into wax. It is extruded in little flakes that the bee picks up with a swipe of her hind legs and then transfers to the mouth to masticate. Then it is communally made into honeycomb.

The precise regularity of honeycomb has been a source of wonder, if not astonishment. There has been debate about how the bees create the shapes of the cells. However, it is known that the workers require head movements to build the comb because if a bee's head is experimentally immobilized (by gluing it to the thorax), she is unable to make comb. Other than that, and after 350 scientific papers about research on honeybee comb and after the publication of a book on honeybee wax, we still don't have "the" full answer to how bees make honeycomb. We do know, though, the chemical composition of the wax, and that it has special properties with regard to temperature effects. Nest temperature is regulated at thirty-five degrees Celsius, and at that temperature the wax is solid but soft enough to be malleable, yet not so soft as to compromise mechanical structure.

A typical honeybee home contains about a 2.5-meter-square comb surface with a total of about one hundred thousand "cells," the hexagonal multipurpose units of the honeycomb. But no one bee makes a whole cell. Each adds onto what another has done, wherever it happens to be, and the result is one of the most beautifully and perfectly crafted structures in the animal kingdom.

Forager bees regurgitate nectar-on-the-way-to-honey into an empty cell or one that already has honey in it, or they scrape off the pollen clumps from their legs and shove that into an empty cell or one that already has pollen in it. Honey production from the dilute nectar is accomplished mostly by climate control of the home. The high temperature that is maintained and the fanning by the bees to circulate air cause water to evaporate from the nectar until it becomes the thick, golden-colored honey, a product that is so concentrated that molds and bacteria cannot grow in it. It is the energy base that drives the hive activity in the summer and is "burned" (metabolized) by the bees in winter to produce heat by shivering.

When by springtime the wax bins are emptied of honey and/ or pollen, the queen inspects the combs and lays an egg into each empty one she finds (up to fifteen hundred per day), and it then becomes a "crib" for a single larval bee. When baby making is no longer on the agenda, the baby cribs revert back to food storage bins. Those cribs made for holding males (also used for honey and pollen storage) are larger than those for the sexually undeveloped females, the workers.

Honeybee combs grow from the top down, and in the hive they are usually placed parallel to each other. The spaces between the combs, or "bee spaces," serve as the crawl space where the bees travel around in the hive and also where they gather. In a hollow tree that is used as a bee home, the lower combs are preferably used as the nursery, and it is here that the dances occur, while the ones higher up in the "attic" of the home are used mainly to store honey.

Our human home-making now is intimately related to the bees', because honeybees are major food providers for us by crop pollination while also yielding us food directly. Many families do (or could) harvest all of their annual sugar needs in the late summer from one beehive, leaving enough to tide the colony over to the next spring. We have established a symbiosis with them. They give us honey and pollination, and we reciprocate by giving them home sites. Like most symbioses, ours started by exploitation.

Access to the honey of bee trees in Europe had involved climbing the tree and hacking away the back of it to expose the hollow with the bee nest. After the honeycombs were taken out, the hole was covered with a door for future access. As the bee tree was periodically visited, the bees were sedated with smoke, new honeycombs were removed, and then the tree was sealed up again. Presumably the forest "beekeepers" left as much of the brood nest intact as possible. In the Bialowieza Reserve in Poland and Russia (possibly the only virgin forest left in all of Europe, where one can still find oaks, basswoods, and pines aged over five hundred years), there are still such centuries-old bee homes in some of the ancient trees once maintained by the now-long-gone forest beekeepers. In east Africa (Tanzania), wild bee management until very recently involved hanging hollowed-out logs horizontally from trees to serve as hives. In Egypt, from the days of the pharaohs, clay cylinders served and still serve the same purpose.

Over most of the world, bees are now handled for agricultural pollination and honey production by providing them with prefabricated homes that accommodate the bees' own extraordinary architecture. Such hives, the 1851 invention of the Rev. Lorenzo Langstroth of Philadelphia, consist of sections called "supers" that are set (like removable floors of a multistory house) one directly on top of the other. This arrangement allows one to enlarge (or contract) a bee home in successive stages as each super gets filled, and then also to disassemble it to examine the contents and remove the honey. But the main innovation of the Langstroth hive is that it channels the bees to build their combs into frames that can be moved, rather than as formerly when the honeycombs were solidly cemented into and throughout their home.

Each super usually has ten removable frames that hold the bee comb that, as in a wild hive, may alternately serve as a brood nursery in the spring after it is emptied of honey, and then again as a storage facility for pollen and honey in late summer and fall. The frames can be pulled up by the beekeeper and the honey extracted by centrifuge, so the combs are used over and over again. Otherwise the combs would be destroyed each time we took the bees' honey, and a typical bee home requires the bees to use up seven kilograms of honey to build about one kilogram of wax — and that does not include the labor of making the honeycomb from the wax. For most of our history before the Langstroth hive, and well into the present in some locations of the world, examining the inside of a bees' nest and getting the honey out required the equivalent of destroying the house to get into the pantry. Giving bees a good home makes it easier for them to make the surplus honey that we take from them for our use, because they need to expend less work in house construction, and because we move their homes to areas where there are many flowers with nectar and pollen.

The Langstroth hive has multiplied the honeybee population to extend to an ever-larger scale, and to distribute it to an ever-larger geographical range. It was made possible by our sensitivity to specific details relevant to the bees. Our relationship to bees extends our already unique stature as the only animal to be symbiotic by giving homes to diverse animals, including cattle, dogs, cats, horses, chickens, turkeys, and geese.

The Langstroth hive gives bees an almost ready-made home and has made honeybee management possible on an industrial scale, as well as accessible to the amateur. Bees are now part of our home environment. They benefit all and strengthen the whole fabric of our existence. Perhaps one can exist without them, but what is the point of existence without seeing them working on flowers, hearing the music of their buzzing, tasting their honey, and drinking their mead? (As if to answer my question, I received an e-mail almost immediately after I had written these words, informing me that the Harvard Microrobotics Lab has a team of computer scientists and engineers making a robobee to pollinate fields of GMO crops, because these robots would be immune to the heavy doses of pesticides and herbicides that Monsanto sells. The robot's reputed "biggest drawback" is that "at the moment" it needs external power; it "flies" on a tether to a power source. Drawback indeed. In anything modeled on nature, power is the central ingredient. All else is window dressing.)

Solitary animals also use mobile homes. Over eight hundred species of hermit crabs (both on land and in the water) avail themselves of the benefits of protection from predators and the elements by carrying discarded snail shells around on their backs. But good snail shells are often hard to come by, and some animals have evolved to construct their own mobile homes, marvels in miniature. Bag moth caterpillars make themselves little portable shelters that they stay in for nearly a year and do not leave until they are adult moths, and in some species they do not even leave them then. Their shelters are made of silk, smooth and white inside but bristling on the outside with all sorts of material, depending on the species. Bag moth homes may resemble a pine cone, or a miniature elongated log cabin with two entrances. The top entrance is where the larval caterpillar can reach out with its head and legs, and at the other end is an exit hole for disposing of wastes. The outside of the bag may be made of any kind of debris the caterpillar encounters: bits of grass, twigs cut to length, or other materials. The construction materials may be affixed either in longitudinal ("log cabin") or perpendicular lines around the length of the bag. At the end of the summer, when the caterpillar has attained its mature size, it closes the front "door" by fastening the front end to a support. It then turns around inside its bag to face the other entrance, and then it pupates. The moth emerges from the pupa the next spring. If it is a female, having no wings and looking like a little sausage, she comes to the back-door entrance on the proper schedule to broadcast sex pheromone and wait for a male to mate with. If successful, she turns around and, with a long extension on the tip of her abdomen, reaches back into the back door and deposits her (up to several hundred) eggs into her just-vacated pupal skin. She may then return inside to die there

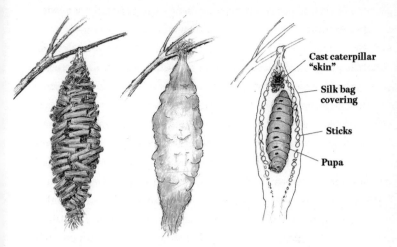

Cast caterpillar
"skin"

Silk bag
covering

Sticks

Pupa

*A home made by a bag moth caterpillar that is (as here) also used by its
pupa. At left is the house made of about one hundred pieces of chewed-to-
size twigs. The caterpillar inside leaves an entrance at the top from where
it reaches out with head and legs and takes its house with it as it feeds. But
after achieving full growth, it seals that entrance and attaches its house
with silk to a solid substrate, such as a twig. It then turns around inside,
molts, and becomes the pupa. The caterpillar's former waste-disposal door
will now provide the eventual exit door when the moth emerges, months
later. This species (collected in Suriname) had an added feature:
it enclosed the whole log structure in silk before pupating (center).*

and leave her body as a loose plug at one end of the sac. In some
species, eggs never leave the female's ovary, bypassing the stage of
shedding the pupal skin.

One bagworm caterpillar I found on an acacia bush in Kenya
had made its home by chewing off the bush's long (about five centimeters), tough sharp spines and using its silk to glue them longitudinally into a tube around itself. It lived in this thorn tube and,
like the bagworm mentioned previously, could extrude its legs and
head from the opening left at the top in order to hold onto the bush,

travel around on it, and feed on the leaves. When threatened, it pulled itself up tight against a branch. Still another bag moth home that I found in Suriname had an added feature to the otherwise similar longitudinal short sticks "log cabin" model. This caterpillar had covered or enclosed the whole "log" structure in a sheet of tough canvaslike silk. I cannot begin to fathom what its function might be, but it must be important because some very fancy acrobatics must be involved in applying it. It is easy to see how a caterpillar can make a silk case surrounding itself. But how does the caterpillar manage to surround something it's already in?

In their defensive mobile homes, these caterpillars can feed in the presence of predators, unlike most others that protect themselves either with poisons or hairlike sharp spines, or with exquisite mechanisms of hiding that greatly constrain when and how they are able to feed.

Despite being strictly aquatic, the larvae of caddisflies (Trichoptera) make almost the same kind of mobile homes that the bag moth caterpillars do. Each caddisfly species has its own specific kinds of building materials and its own often very unique way of assembling them to build its home. In some, the materials used, such as cut pieces of live plants, dead twigs, pebbles, flattened stones, crustacean shell pieces, or pieces of leaf, are assembled in random (radial) tubes. But one species, *Lepidostoma hirtum*, makes a square case by silking rectangular leaf panels into four rows to make a four-sided tubelike box. These artistic creations have also inspired art; the French artist Hubert Duprat "produces" (sort of) variations of the naturally made caddisfly cases when he relocates larvae to aquaria in his studio and provides them with flakes of gold and precious stones, and they then build their cases entirely of these materials.

In Alder Brook next to my home in Maine live at least five common caddisfly species, each making a very specific "case." There the

Samples of the "cases" made and used as portable homes by the larvae of different species of caddisflies. The four cases on the left are mobile and found in still water. The three on the right, found in flowing water, are fixed in place and also serve as food traps.

larvae have a large choice of available materials, and they exert their specific preferences. One of these species constructs a flattened case from sand grains that rests on the bottom of sandy stream areas. Another uses small pebbles; a third, long thin pieces of bark glued together lengthwise into a round tube home; and a fourth makes its house by cutting grass into short lengths and silking them crosswise rather than longitudinally like those of the bagworm moths described earlier. In each instance, the house mimics some feature of its surroundings and allows the occupant to move around freely. Because of its home, the caddisfly lives where it would oth-

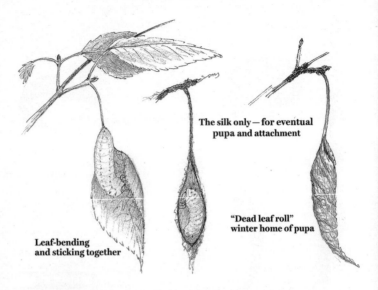

The silk only — for eventual
pupa and attachment

"Dead leaf roll"
winter home of pupa

Leaf-bending
and sticking together

A promethea moth caterpillar (Collosama promethea) *making a home, the cocoon, for its eventual pupa. The caterpillar begins by attaching bands of tough silk that connect the petiole of a leaf to the twig. The leaf is rolled up and the inside is lined with a thick, tough layer of silk. During the winter the cocoon remains attached to the twig and mimics a dried leaf.*

erwise, within minutes, probably be eaten by this brook's trout and minnows.

One might suppose that the ultimate home is one that you do not need to either leave or travel to. And indeed, the larvae of one caddisfly species, *Neureclipsis bimaculata,* achieves both. But its strategy of making the home a food trap rather than going to the food requires a different diet and habitat, namely, flowing water containing plankton. This caddisfly larva extends and flares the entrance of its house with silk, so it acts like a seine. This home entrance faces the water current, and the larva stays in its home at the bottom of the net to catch prey that drifts in.

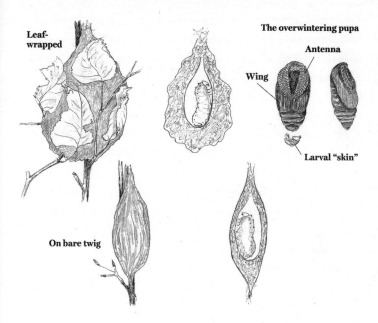

A cecropia moth caterpillar (Hyalophara cecropia) *may pull several leaves together (top), make a loose papery cocoon, and fill it with a thick layer of fluffy, cottony silk and a tough second shell inside. An exit hole is left for the eventual moth to escape in the spring. The caterpillar may also attach the cocoon without wrapping it to adjacent leaves (bottom), by simply attaching it to the sides of a twig. The caterpillar (center) finishes making the inside layer of its cocoon before it molts to pupa.*

Some of the more obvious homes made by larvae, and most easily taken for granted, are the cocoons that house the pupae of moths. They serve as a safe place for the pupa, the immobile stage of the animal between the caterpillar and the adult stages. The majority of moths in the Northern Hemisphere spend about nine months in this stage, during which they are helpless and depend for survival almost entirely on their home. In some moth species, for example, sphinx and owlet moths, the larvae burrow underground and create a small cavity there that serves the same function as a cocoon. Many other moths create homes out of silk and stay aboveground. These may be fortresslike solid structures and may employ "tricks" that that hide the otherwise helpless pupa within. The silk used to make the cocoon of one species, *Bombyx mori*, is well known as the raw material with which we produce fine fabric.

Of locally common species of "silk moths" in North America, I'm most familiar with the luna, promethea, polyphemus, and cecropia moths. The luna and polyphemus caterpillars prepare for their nine-month pupal stage by wrapping leaves about themselves and then within that cavity surrounding themselves with a layer of silk. The silk layer is flimsy and papery, but it holds the leaves together, serving as camouflage. The cocoon falls to the ground when the tree sheds its leaves, and then it becomes buried under the autumn leaf fall and overwinters under the snow. The promethea moth caterpillar also starts with a leaf wrap but makes a much harder, more leathery shell-like cocoon that is much smaller, for a tighter fit around the pupa. The promethea caterpillar usually wraps a single leaf, and at a sock-top-like exit hole it creates an extension of the home in the form of a thick strap of silk laid down along the petiole of the leaf and solidly wraps that around the branch. As a result, these cocoons remain attached to the twigs when the tree sheds its leaves. They look like the old wrinkled leaves that sometimes remain on trees all winter. The cecropia moth's strategy of a safe home appears to rely on a psychological trick on potential predators such as squirrels, mice, and birds that prey on pupae. It is a very large moth, with a pupa that could be (and is!) quite conspicuous. Indeed, the cocoon looks balloonlike. It has an outer shell and a second inner shell within which the pupa resides. But there is a wide empty space in between. Presumably a predator that penetrates the outer shell would find what appears to be empty space after penetrating it, not realizing there is a second room beyond the first.

Although the preceding cases exemplify homes made by solitary juvenile animals that serve as protection for themselves or a future stage of their life cycle, other solitary animals build homes as adults for their multiple larvae, which they then provision with food. In solitary wasps, the larval food is usually paralyzed insect and spider prey that is carried into either a natural cavity or one that is excavated to become the nest. My father told me of an unusual one in Angola, Africa. He had, as always, hung up his shotgun outside the tent when he noticed one day that the barrel was partially filled with bits of charcoal. Discharging the gun with a plugged barrel might have resulted in an explosion. He suspected one of his African helpers was up to no good but then discovered that the culprit was a blue wasp, busily carrying in the charcoal. It also brought pill bugs (a terrestrial crustacean). The wasp separated different individual pill bugs with layers of charcoal, and thus each larva had its own compartment with food. Potter (also called mason) wasps (genus *Eumenes*) may burrow in wood or use containers such as those created by a wood-boring beetle larva, or they make a narrow-necked round pot from earth and regurgitated fluid. The form and mechanism of making such pots could have been a model for those made by Native Americans.

Some bees, called mason bees, genus *Osmia,* also make similar clay pots. But these are provisioned with pollen as a protein source for the growing larvae, instead of paralyzed prey. Other mason bees nest in preexisting tubes. Such bees are used for crop pollination — they are provided with homes that attract them. These are simply blocks of wood drilled full of holes and set near a pollen source, such as an alfalfa field or a fruit orchard. Some of these bees and their homes are now commercially available, but I simply provide a debarked old pine log full of holes made by larvae of the long-horned beetle pine borers. Give them homes, and they come.

.   .   .

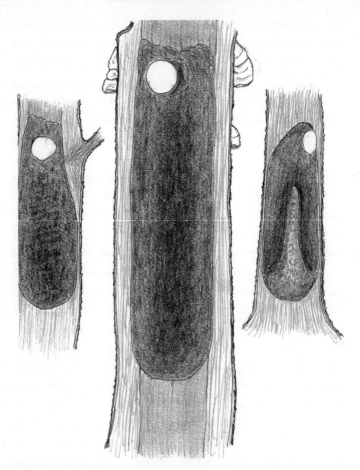

Diagrammatical views through cuts of trees to show nest cavities of a yellow-bellied sapsucker (center), and two of black-capped chickadees. Sapsuckers prefer to use poplar trees whose centers are softened by the hoof fungus. They do not reuse their nest holes, but other birds may use them, such as chickadees. Chickadees may also hammer out their own nest holes. The two shown here were in dead, partially decayed quaking aspens. One nest hole (at right) was left unused after the birds excavated all around a hard core and then abandoned their effort. Woodpeckers do not put a nest into their nest holes, but the various species of birds that use such abandoned cavities do.

The homes of birds function in the majority of cases for one-time use to hold the eggs, incubate them, and rear a clutch of young. But some serve also as dormitories or shelters in the winter. Bowerbirds produce a variation of these homes as display objects to attract a mate. Some woodpeckers, such as the downy and hairy woodpeckers in North America, hammer out cavities in both spring and summer, the first to rear young and the second as a shelter in which to overnight.

The nests of mammals parallel those of birds but usually serve more than one function at the same time. Those of deer or white-footed mice, genus *Peromyscus*, are used to raise young in the spring and summer, as a granary for storing seeds in the fall, and as a place to huddle with others to keep warm in the winter. One mid-October as I was showing students a season's discarded catbird nest in a bog, we saw two of these mice jump out. I was surprised, because a catbird nest is a cuplike structure of roots and rough bark on a platform of twigs. It would not be a shelter for a mouse. But a closer inspection revealed that this pair of mice had done some renovations. They had separated the nest lining of roots from the base of twigs and had then stuffed the space created in between with the white fluffy down of milkweed seeds. I discovered another of these mouse nests in the glove compartment of my Toyota pickup truck, which was parked by the house in midwinter. I had opened the compartment to search for my registration, only to find that all the papers in there had been shredded to make an insulated mouse nest. Still a third deer mouse nest was in the (unused) paper delivery box on a pole next to a mailbox. Another one, on top of a chestnut-sided warbler nest, was filled to the brim with black cherry seeds and covered with the down of *Clematis* seeds.

The home-making of birds and mammals is separated by two main life strategies: those of the "altricial" or helpless young, such as most songbirds' that are born blind and helpless except for being able to beg for and take in food and excrete waste; and those of the "precocial" young, such as those of ducks and chickens that are mobile from birth and don't use a nest. In the precocial strategy, the main solutions that help reduce predation are behavioral, such as guarding by the male mate (in the case of large birds such as cranes,

Instead of excavating cavities in trees (or the ground), some birds enclose space by the use of mortar. The nests made inside these structures consist of feathers and fine plant material. Those of cliff swallows, illustrated here, are often next to each other in colonies where one nest may serve as an attachment point for the next.

swans, geese), deceiving a potential predator by feigning injury and luring it away from the young, and/or hiding responses of the young (in the case of grouse and sandpipers). Having a nest, however, has many potential advantages. First, it provides other options beyond just hiding. They include placement of the young into inaccessible

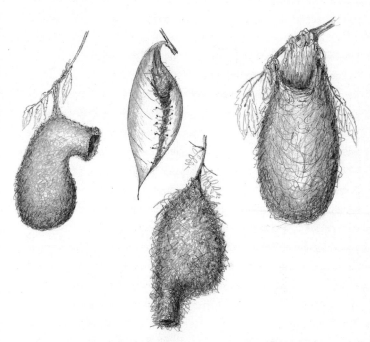

Other birds make relatively inaccessible nests by weaving them and hanging them from the tips of branches. Here, from left to right, are four examples: The Eurasian penduline titmouse, Remiz pendulinus, *makes a feltlike bag nest. The Asian tailorbird,* Orthotomus sutorius, *pokes holes into a large leaf and, using its bill as a needle, threads fiber through the holes to bind the two leaves or leaf ends together. The African oryx weaverbird,* Euplectes orix, *weaves grasses. North American Baltimore orioles,* Icterus galbula, *in New England commonly extract fibers from year-old dead milkweed plants to weave with.*

places, such as over water, high on a cliff, onto the ceiling deep in a cave, at the end of a thin twig, and in uncountable possible hiding places — provided the adults can have access to feed their young. Very specific building materials and construction techniques are required.

The "inventions" that we see in birds are seemingly endless and sometimes border on the bizarre, but they are always beautiful and sometimes astounding. Simple nests have the virtue of being "invisible," and thus a mere scrape on the ground may suffice. Others, however, involve skills to construct that challenge our imagination. My own favorite examples of marvels of construction are the hanging nests, those that have evolved the furthest from scrapes on the ground. Consider the tailorbird, *Orthotomus sutorius*, which makes holes in leaves on a plant and then finds fiber to poke through these holes and stitch the leaves together as a surgeon might suture skin. The bird then builds its cozy nest inside the leaf bag it has created. A variety of swallows create a similar home cavity within which they then build a nest, only they follow different options (depending on the species), such as finding an existing cavity such as one made by a woodpecker, excavating one themselves in the ground, or, as most do, mortaring one together out of clayey mud. Nest cavities are also created on the designs of clever baskets. Many weaverbirds expertly weave fibers of grass into retorts that are hung from the tips of thin branches where they are out of reach of many predators. They have a long, narrow tube at the bottom that acts as a secondary door which restricts access as well. On the same principle, there are also designs, such as in one titmouse, where there is a door flap that is closed when the bird leaves the home, and still another has a false entrance. Nest location is important. Hardly any space would be safer than the ceiling in a cave, and some swifts have achieved the occupancy of that perhaps safest of niches. They have built nests out of their own saliva by adding one bit at a time, which hardens and extends a shelflike projection on which they can deposit an egg. Many swallows achieve somewhat of the equivalent by using wet mortar that they collect.

A fortress, if it is impregnable, does not need to be hidden. Cavities, especially those in trees, are one of the most common ready-made fortresses. There are few major bird groups in which some members do *not* avail themselves of this resource. The problem is that it is often a limited resource, and depending on this scarce resource may severely limit reproductive opportunity. Therefore, the ultimate in home-making for hole nesters is to excavate their own home, to their own specifications. Woodpeckers, perhaps because they are pre-adapted for it because of their foraging for wood-boring grubs, are the experts at it.

Woodpeckers excavate their own home cavities out of solid wood, with the entrance reduced to a minimum size that excludes almost all predators larger than they are. Small predators that might be able to squeeze in can likely be repelled with the same tool used to make the home, the sharp and powerful bill. Larger predators that can't enter might still be able to reach in through the hole, but if the cavity is made deep enough, they can't reach the eggs and young on the bottom. Woodpeckers may expend what appears to be extraordinary effort to make a home, but the effort is worth it, because they suffer virtually no nest predation.

Woodpeckers make no secret of where their homes are. In June when they have young, I can locate a woodpecker nest from within a hundred meters, whereas I may step over any of a variety of other forest bird nests without noticing them. A woodpecker comes and goes to its nest at frequent intervals, and the young in the nest make a nonstop din that increases in volume whenever they sense something landing on their home tree. All other forest bird babies stay silent until a parent is almost directly at the nest, and then the babies peep only faintly and only the moment the parent arrives with food. They shut up the moment it leaves. There is never (in most species) a single fecal droplet in or near the nest due to a complex interactive behavior between the parents and their offspring. In contrast, with the woodpecker nest, after the older young start to monopolize the nest entrance, feces accumulate and are simply trodden down into debris that collects on the floor. There is, it appears, a downside to almost everything, and each solution is a balance of trade-offs.

Similar scenarios of homing apply in mammals as well, and the two divergent strategies of altricial versus precocial young are exhibited in closely related species. Those of the hares and rabbits, Leporidae, and apes and humans, Hominidae, are the most interesting and instructive examples.

Hares and rabbits look very similar to our eyes but have vastly different life and home strategies. Hares, *Lepus,* such as jackrabbits and varying or snowshoe hares, bear their young into a simple bare depression on the ground. They have no protective home but, like precocial birds, are born with insulating and camouflaging covering. Like baby sandpipers, ducks, chickens, deer, and antelope, they are able to run from birth. Rabbits, on the other hand, are born blind and naked into a fur-lined nest (the hair the mother has plucked from her body). One species, the European rabbit, *Oryctolagus cuniculus,* is the ancestor from which all domesticated rabbits are derived, and in the wild, this species digs networks of tunnels to create what is called a "warren." One might suppose that living in an excavated cavity is safer than having the young deposited in the open on the ground. It could be, but an earthen tunnel is not impregnable. Some predators have evolved to enter tunnels. A rabbit warren must have many potential escape exits. But it has an additional design feature — a nursery chamber with the nest and babies in a side tunnel that is without an exit. This is not a design flaw but a "clever" safety feature, since the helpless young could not escape in any case. Rather than allowing a predator that searches within a warren to reach the nest from two directions, there is only one. Furthermore, this entrance is plugged most of the time, and the doe (female rabbit) reduces the time window when the tunnel to the nest is open to a bare minimum — the three or four minutes per day, at almost exactly the same time every twenty-four hours, that she visits to nurse them. The young learn when to expect her and get ready by uncovering themselves to be ready to nurse the moment she comes in. After having nursed her babies, she leaves immediately and plugs the nest entrance behind her.

It is probably not incidental that this species of rabbit is social as well. (In contrast, cottontail rabbits of the Americas, genus *Sylvilagus,* instead have extremely well-camouflaged ground nests.) An excavated home with many tunnels represents considerable investment, and it is an asset rather than a liability to have social tolerance for helpers with an equal stake to make, enlarge, and maintain it (as also in beavers).

Few if any home-making feats are as impressive for sheer size, sophistication, and ecological impact as beavers'. Beavers don't just make their nest, called a "lodge," but also create their own specific habitat all around it by clearing trees. Felling trees up to half a meter in diameter with their teeth may sometimes require days, and the rewards from the work come only from the branches near the top. The twigs' bark is their main food, and the peeled remains become one of the main materials for making the lodge, as well as the dam that then creates the moat around the lodge, the safe haven over the winter and where their kits are born.

The dam, which creates a pond that serves as a moat, permits the home entrance to be underwater and keeps out unwanted visitors. In the winter this home has a roof and walls that freeze as solid as concrete. As long as the dam holds the water and the frost hardens the lodge, the beavers' home is impregnable to bears and wolves.

This lodge is no ordinary rodent nest, as John Coulter undoubtedly surmised. (Coulter, considered the "first mountain man," lived for months alone in the wild and was the first white man to see the Yellowstone area and the Grand Tetons. He once escaped from pursuing Blackfeet warriors in the Yellowstone in 1809 by hiding in a beaver lodge.)

Beavers' engineering skills are most evident in their dam building, which may impound acres of pond and raises water levels one or more meters to make the safety in their lodges possible. The results of their collective work, often performed over generations and presumably hundreds of years, can be stunning. One beaver dam in Wyoming, though only ten meters long, was five meters high. Another on the Jefferson River in Montana measured 650 meters long, and one in Wood Buffalo National Park in Alberta was slightly more than 850 meters long. Dams break not infrequently, and their upkeep, repair, and replacement are almost year-round tasks. In the last spring rains, one sixty-meter dam backing up our pond of several acres was breached. The water rushed out, leaving a large mudflat. But within one week the beavers had repaired the dam. Close to that same time, in the nearby town of Adamant, Vermont, a beaver dam broke and seven houses had to be evacuated because of the flood. Beaver home-building reminds us that home doesn't stop at the door of the dwelling. It includes the area from where we secure our resources to live on. The larger the area of their pond, the farther the beavers can range to reach more food in safety. The more the animal can do to make its place habitable, the larger the range of habitats it can use.

A typical beaver colony starts out with an adult male and female, who have several kits in early spring that grow up by fall to help the parents in dam and lodge building and in harvesting the food stores required to overwinter. By the following spring the parents have another clutch of pups, and the colony then contains perhaps up to seven or eight individuals. The previous set of young then effectively helps the second. But by the following spring, before the pair has its next set of pups, they expel the two-year-olds if they don't leave on their own, and thus the colony size stays usually below seven or eight individuals (although a theoretical fourteen is possible). Those that leave home face predators, which traditionally were wolves and coyotes, but judging from the dead ones I see every spring along roadsides, it's now cars.

The better a home the animal can make, the more needs are satisfied directly in its home area and the less it needs to roam. We are dependent on our homes, yet we often simply take them for granted — until we are left without one. A trip into the wilds of Suriname gave a party of us from New England homes a demonstration of where the limits of adventure abut our abilities of home-making.

HOME-MAKING IN SURINAME

AFTER JUST BARELY FINISHING MY LAST-MINUTE TASKS before a trip I'd signed on to as a replacement only a week earlier, I boarded an airplane with one small piece of luggage and arrived after midnight at Suriname's Paramaribo-Zanderij International Airport. There were no lights in any direction. The jungle was dark. I was approached by a cabdriver who, after a one-hour drive at high speed along a straight narrow road, dropped me off at a hotel where I had been instructed to appear. It was near midnight by then. I stepped onto a silent unlit street into warm humid air. And as I contemplated what to do next, a man walked toward me. He gestured for me to follow him, and with little more ado he led me off the street and unlocked the hotel door and led me through a short dark hallway to a room. Thankful and tired, I stepped into it. My eyes adjusted to the gloom and — I beheld a woman with one bare leg protruding from under a sheet on one of two narrow beds. I took the other, waiting to make introductions in the morning.

Early the next day I met up with not only my roommate, the presumed cook, but also the other two members of our Peabody Museum of Yale University team coming here to collect birds in some pristine mountains where no humans had reputedly ever been before. I had as a college student collected birds in Africa for the same museum and therefore eagerly accepted the invitation, perhaps to relive a time when I was still a wanderer without a home.

We taxied to the "Hi-Jet" Helicopter Services. It's the only place in Suriname where helicopters can be chartered, and three of them were parked there. We met Glenn, a former bush pilot from Alaska, who had had military training and would fly us in. I was curious to find out how we would land in the jungle. He confided that he didn't know but offered: "I guess we'll have to look for our own LZ" — Vietnam-speak for "landing zone." This was not Vietnam, though, and as far as I knew there are *no* open spaces unclaimed by trees in an undisturbed tropical jungle, especially in a vast expanse of mostly unnamed mountains. I was thinking of my home now, the New England forest, and my hill in Maine, where keeping my clearing open is an annual chore. However, our leader, Kristof, had mentioned that a herpetologist had at one time flown over the Wilhelmina range where we were going, looked down, and seen what he thought was a bare spot.

Suriname is a South American country of about 262,000 square kilometers that harbors the largest unexplored forest area in the world, and it is adjacent to the great Brazilian rainforest. Most of the people there live along the northern (Atlantic) coast, but our plan would land us in the interior, somewhere among the countless unnamed peaks in the sea of forest surrounding them. Tall mountains are like islands for birds that are at home there. Stranded there, they may evolve to become new varieties and perhaps species, similar to how Darwin's finches differentiated into a variety of species stranded on the different islands of the Galápagos. That was why we were going there: to try to discover a new species of bird.

. . .

All systems were finally go on the third day, after extensive bureau-
cratic discussions over which and how many "specimens" we would
be allowed to take, and whether or not if we took two blood samples and a feather from the same bird we skinned and stuffed that
was one specimen, or four. As far as I know, all of the legal issues
were not resolved. But we were assigned to take along two men
from the Suriname Game Department. One, named Sami, was an
elderly bald man of ebony complexion wielding a large bush knife,
the necessary tool for advancing through jungle. His steady dis-
position won our immediate trust. The other, much younger, was
more of the hip-hop generation and probably more at home in
town than in the jungle. Our cheerful copilot, Lorenzo, a Surinam-
ese of part Amerindian extraction, clad in a bright orange jump
suit, beckoned us to load up, and he made sure we were strapped
and packed in like sardines in a can. He gave the thumbs-up to
Glenn, who then gunned the engine, and we lifted up in the roar
of the twirling blades. Only a few seconds later, we were already
zooming over the treetops at perhaps some 160 kilometers per
hour. And so for the next hour we flew on over unbroken forest
that was, after a few minutes, devoid of human signs. The tree
crowns were closely interstitched in a quilt of shapes — flat tops,
bulges, globes — and colors — mostly a study in greens but with
tints of yellow, blue, and brown. Occasionally we passed a flowering tree that would stand out like a light — perhaps one in bright
white, pink, purple, yellow, orange, and occasionally also rich blue.
I peered down to one dead tree that cradled a huge raptor nest in a
crotch, perhaps that of a harpy eagle. A flock of about a half-dozen
scarlet macaws passed directly below us. It was an entrancing ride
over the mostly green carpet, and it seemed to go on smoothly,
tranquilly, and mesmerizingly forever — that is, at least until we
got into the interior, where we started to see clouds, and then hit
thick fog. There had, so far, not been a single open space between

the trees. Not a speck of ground was visible the whole time. Not one LZ in sight. Glenn tried to skirt the cloud banks because the helicopter was not equipped with radar to detect what they might hide, such as possibly the precipitous mountains that we were approaching.

The fog banks got larger and became more frequent, but turning back and forfeiting a nineteen-thousand-dollar ride was a hard option. We had no way of knowing if we could find clear skies on another day and, I presumed, again that much cash. We still had a long way to go, but for now, still in the lowlands, there was a cleared area, possibly a small airstrip. We turned to find it and landed there to wait out the weather. Here, I saw puffbirds digging their nest holes in the sandy, undisturbed grassy runway, next to a hand-painted sign reading "Vlieguela Paesoegroenoe." There were no aircraft here except ours. In a bush near where we landed, a dark sparrow-size bird hopped endlessly back and forth in a monotonous dance. We walked to a nearby river and saw deep dark water flow by. A man in a dugout canoe paddled by, and one after another, Uraniidae — day-active moths that superficially look like swallowtail butterflies — crossed over the water, each of them flying in almost precisely the same direction. It was a migration of some sort, and it didn't make any sense to me. Where were these moths going, and why?

After a wait of a couple of hours, Lorenzo had us load up and we lifted off, but within several minutes Glenn turned around and we came back to the same spot; it was still too risky to fly. We again went to the river and took pictures. A group of kids appeared and posed with each of us for photos at the helicopter. We were clearly still in civilization. We still had a long way to go.

Eventually Glenn decided to give it yet another try. This time we flew on farther, yet dense fog banks reappeared. He tried to fly around them. Steep rocky cliffs suddenly emerged in front of us.

Glenn swerved the machine, making a hairpin turn to avoid being smashed by cliffs or trapped by the fog. After several such hairpin maneuvers we sat clutching our seats, white-knuckled, sweat dripping from our faces. Those who could turn pale did. I smelled vomit.

It was more difficult than ever for me to imagine how there could be an LZ among steep, densely jungled mountains. But after skimming close over the peak of one and then down its steep slope on the other side, we finally spied a black-brown patch that could only be rock. Glenn instantly swooped down and tilted the machine so we could look more closely. We then turned a tight circle so Lorenzo could get a close look from the open door: he gave the thumbs-up — "It's good!" There was enough space that the rotor would not hit trees. Gas was running low — there was not enough left to search further and also allow the copter to fly back.

We landed on the rock shelf, jumped out, and unloaded. In a couple of minutes, Lorenzo had jumped back on and the rotors sprang back to full force, the helicopter leaping over the lip of our rocky place, and then disappearing as a speck between the craggy peaks that were still half shrouded in clouds. The sound of the engine faded, and we were alone.

We had landed on ancient black-weathered sandstone strewn with flat loose rocks between which grew some grass, small aloes, and orchids. A bare ledge rose steeply to a forested ridge above us. A three-hundred-meter drop-off into a forested valley yawned below us, from where a river roared. Dark tree-clad mountains on the other side of the drop-off showed two huge yellow scars of steep ground where a forest had recently been pulled down by a landslide. Strange birds? One sounded like a saw ripping through metal, unlike anything I'd ever heard. Clouds started swirling down from the peaks, contracting our world. It was then, very soon, that we felt the first raindrops.

This, our new home for the next three weeks, we identified by the GPS coordinates of 3 degrees and 45.02 minutes north, and 56 degrees and 31.20 minutes west. We were in the Sipaliwini District of the Wilhelmina Mountains, at an elevation of 905 meters. These numbers, and only they (if Glenn and Lorenzo had secured them as we presumed), would mitigate the chances of our becoming permanent jungle residents, because our satellite phone, the only link to our homes, would become inoperative by the next day. But for now, the soon-to-arrive downpour had consequences for our trying to make ourselves a home before night.

A level place with soil for pegging down tents had to be found; tree stems were needed for poles; frames had to be set up for tarps to cover the tents. We needed not only poles, but also firewood and drinking water. None of this could have been even considered during our need, and then haste, to land. But we were lucky to find a patch of thin soil in a tangle of growth at the edge of our rock shelf. This was mangrove, and the trees' gnarly brushy branches sent down a phalanx of solid roots into the ground. A thick, thin-stemmed tangle of bamboo filled in every square meter to create an almost solid wall. A saving grace was that a few thin trees had penetrated the thicket, and we used them to erect a framework that we lashed together with twine. Thankfully someone had thought ahead to bring this essential but easy-to-overlook item. We spread our tarpaulin over the frame to make a rough shelter, cleared the ground as well as we could of bamboo, and set up both our sleeping tents and a work tent. Major requirements for our temporary home had been met, except for water — but we soon got more than we wished for.

The rain picked up as darkness set in, and then we heard frogs. One that was stationed near us made a resounding, hoarse, muffled, and not very welcoming *moof.* Or was it a *move*? It repeated this call endlessly at about one-second intervals. The sound echoed around the hillsides and was answered by others. And then the heavens opened in a deluge the likes of which I had never before witnessed. The pounding, pummeling rain roared onto the tarp, and within minutes water came rushing down from the slope above us, straight into and through our campsite. We frantically dug trenches with our machetes to try to divert some of the flood but could not prevent it from washing under and around our tents. A wet and uncomfortable first night ensued, one not unlike others that followed, except for the fact that the aforementioned satellite phone was eliminated (despite having been wrapped in plastic as a precaution against wetting) as a possible connection to home and loved ones, who would be expecting to hear from us.

At dawn that first morning I crawled out from under my rumpled wet sheet, imprinted with the roots, rocks, and sharp tops of bush stumps we had not cut off quite flush against the ground. The mountains all around were shrouded in drifting veils of fog that wafted slowly up from the valleys, with tall trees standing in the fog on the ridge top above us. One ridge after another extended into the distance. Water roared over falls and between the precipitously steep mountains in rivers and rivulets, presumably eroding the gullies ever deeper, as they have for millions of years. These are the oldest rocks on Earth, Precambrian. Fog in the deeper valleys drifted up slowly like the smoke from a thousand fires — and thinned out and joined the clouds. A constant din of crickets was muted by the roar of the water and punctuated by strange grunts, whistles, trills, and screams of birds waking up to start a new day. The next day we found a coral snake slithering out from under a tent. But it wasn't only snakes we worried about. Flies could be more dangerous, because you can't always see them, and identification is difficult because the dangerous, disease-carrying ones don't advertise themselves with bright colors. Even before I could hope on the first morning for the hot coffee that wasn't, I learned from one of the crew members for the first time about a fly that injects a parasite that devours the host's cartilage, starting with the nose, moving on to the ears, and then, as it multiplies, also the cartilage in the joints — it's a painful death. Naturally, we tried to keep the work tent flap zipper tightly closed to keep out insects. But after three days of constant zipping and unzipping, the tent flap zipper zipped no more.

Sami, chopping brush with his machete to enlarge our space, found a big green snake, one he thought was a harmless "python" but later turned out to be an unusually aggressive and deadly viper. Kristof managed to lift it with a stick, grab it behind the neck, and inject it with Nembutal, at which point it opened its mouth and revealed long and pointy white fangs. Sami also made the first trail out of our enclosure, almost prison, on the rocks and found a nest of honeybees — the recently imported so-called killer bees — in a hollow tree — or rather, they found him.

The bees later also found the rest of us; they started entering our work tent and buzzing around as Kristof and I were skinning and stuffing birds after our morning hunts. With no sweets nearby, they were not after food. But, knowing what bees like for a home site, I suspected that they were scouts searching for a potential nest site. If they found our sheltered spot, such as a cozy corner under our tent flap, a suitable nest site, they might consider our temporary home a suitable permanent one of their own and bring their swarm. Thinking of my friend and colleague Tom Seeley's work, I knew it was essential that they not get a quorum for a collective positive swarm decision, and I made a special effort to squash each and every one that got into our tent. I killed about a dozen.

Bird nests were high on our list of desirables, aside from their makers, and already on the second day we found some. Close to our tent at the edge of the clearing a wren had attached its nest to the tip of a long slender branch. This nest was clearly visible, but probably inconvenient for a rodent or a snake to reach. Another was hidden on the ground in grass by our bare ledge at camp. I showed it to Kristof, an ornithologist who is knowledgeable about South American birds and their nests, and he then looked around and found another one just like it within five meters of the first. This one, probably from the second clutch of the same bird, contained one white egg. We subjected the egg to the water test; it floated, indicating that it had been incubated, so this was indeed a one-egg clutch. Tropical birds face high nest predation by snakes, which find the nests by watching birds' activity of returning and leaving. So, the fewer the young, the fewer the nest visits by the parents, which increases the possibility of raising the young to fledging. We set up mist nets and caught the bird. It was brown with a strikingly bright scarlet iris, identified as the rufous-tailed tyrant, *Knipolegus poecilurus*. This was the first bird we caught at our new home base, and it was new for Suriname. Furthermore, the nest had never been recorded even from other parts of the species' range in South America.

In this forest the most conspicuous nests were those of termites. Some were a half-meter long and nearly as wide. They looked like grotesque black growths on the trees' trunks. I cut into one of these termite homes with my machete and was surprised that the material it was made of felt and looked like plastic. It was obviously water resistant, since it does not dissolve in the constant drenching to which it is subjected. Like all termite nests, it likely contained the makers' feces as a binder. Might its composition hold a secret to a new and less toxic plastic substitute?

Frogs, from the sound of it, were very much at home here. But of all of the various animals that are home builders, frogs are not distinguished. Instead, they have evolved ways of parenting that for the most part do not require nest making. Akin to primates and some spiders, they have invented ways of carrying their offspring around with them (on their backs or, in the case of one species, in their mouths).

All of the frogs I know back home simply deposit their eggs into ready-made pools. But here on our mountain, just above our campsite on the rocks, there were no natural pools. Even if there were a rock cranny that might hold some water where a frog might then lay its eggs and hatch tadpoles, all would be washed down the rock face by the water torrents that came through regularly at night. If not that, then the hot blazing sun would evaporate all moisture in the daytime. This rock face seemed the most unlikely frog habitat that I could possibly imagine, especially in this otherwise watery world. Would or could a frog really make a home here?

The question was rhetorical, because right there on the slope we found what looked like a mug-size gob of white foam jelly stuck to the rock and anchored to some grass coming out of a crack on the

Suriname frog guarding eggs and already relatively large
(four-centimeter) tadpoles in its pool, apparently created by bubbles

rock face. Looking closer, I saw a small frog half submerged in it. A frog in a nest of bubbles? It was no hallucination, though — because looking around I found a second one near it.

After I kept track of the frogs' nests for sixteen days in succession, it was clear that these frogs had accomplished something elegant. The frog homes looked like the very common and familiar little foam bubbles ("cuckoo spit") where the larvae of leafhoppers, genus *Cicadella,* live. Each leafhopper larva makes its own home from secretions of glands in its abdomen that it whips into a froth using an air pump (derived from its breathing apparatus). The bubbly froth protects the larva until it matures to the adult stage, when it looks like a miniature cicada, though usually a very colorful one such as bright green, blue, white, yellow, or black and decorated with red spots and stripes.

This frog's foam made a small dam. With time I noticed that the white froth gradually changed to a consistency more like jelly, and under it I found liquid. I do not know how this frog had made the froth to make a dam. Possibly it did something like a Javanese frog, in which species the froth making is associated with mating; as the female lays eggs, the male is on top of her and fertilizes them; they both dip their hind feet in mucus that is being excreted along with the eggs, and with their feet they then whip up the foam that surrounds the eggs as they are excreted. In the Suriname frogs, the foam formed a ball that was stuck to leaves or grasses, and later the foam inside of the ball liquefied, and the larvae were then enclosed in a miniature pool.

The frog in attendance seemed so obsessed with staying in its home that it allowed me to pick it up and take its picture while it was in my hand. It went right back to its original place and stayed there when I set it back down. On looking closer, I saw on this first day (April 2) frog's spawn (eggs) within and under the jelly. On the morning after a torrential pounding, with gushing downpours during the night of April 17, the little artificial pool contained another batch of about two hundred more fresh eggs, in addition to a squirming mass of four-centimeter-long tadpoles, showing a surprising degree of home permanence.

The frog had made a sufficient "pond" for itself on a steep rock ledge where the sun beat down in the daytime, and where the rushing water in the nightly rains washed the ledges clean. Admittedly, many insects and some birds perform similar feats of home building using their feces, their spittle, with mud and sticks and rocks, but here in this inhospitable place the frog had made a home, where it had grown its tadpoles, with little more than the secretions from its cloaca. That's elegance.

The Suriname frog had a dark warty back, contrasting with brilliant crimson patches separated by dark stripes that it exposed when it extended its hind legs. Maybe I should have "collected" it, because I now wish I knew its name. My photographs were examined by Dr. Rafael Ernst, who had worked on the amphibians of Suriname and is curator of herpetology at the Museum für Tierkunde in Dresden, Germany. He informed me that it could have been *Leptodactylus rugosus*, a species restricted to the high-altitude ranges of the Guiana Shield. "However, the bright red coloration is unusual," and he suspected it was a new species.

On some nights, after our tedious work sitting at a camp table in the confines of the tent skinning and preparing birds for the Yale Peabody Museum collections, we took the opportunity to catch moths. James Prozek, one of our group, had brought along a "black light," a common entomological tool for attracting insects at night. We wanted to capture the almost mythical "white witch," *Thysania agrippina*, a moth distinguished for having the largest wingspread of any moth or butterfly in the world. We set up the light at the edge of the drop-off, facing the valley and a mountain beyond, with a white sheet behind it for moths to land on. Night-flying insects use the moon as a reference for "homing" orientation. Although they may not have homes as such, they still need to fly in consistently uniform directions to achieve distance in finding mates and food plants. The adaptation to use a bright natural light to orient to dooms them when they encounter an "artificial" light that is near, because they (like night-flying migrant birds) get caught by it, zigzagging or circling to and into it. White witches flying on an overcast night in the valley beyond our light would, we hoped, be misled by our beacon. And, to our delight, several of them came fluttering in. But my favorites at our light were the bycatch sphinx moths, the nocturnally flying analogues of hummingbirds. Some species are heavier than the witch, and among moths they may be unsurpassed in the beauty and diversity of color patterns in different species. I ended up making a collection of them to bring back home.

It was not until long after coming back home, when I tried to identify some of the species, that I came upon the work of an explorer in Suriname, Maria Sibylla Merian. Merian was a native of Frankfurt, Germany, who was sponsored in 1699 by the city of Amsterdam to travel to Suriname, then a Dutch colony, to study and draw insects. She apparently had a fascination with sphinx moths. Of the nearly two dozen species I collected at our light, I later learned that one, *Cocytius antaeus,* had been illustrated by her about 310 years earlier, along with its caterpillar and its food plant. In her book about the insects of Suriname, *Metamorphosis Insectorum Surinamensium,* published in 1705, Merian illustrated not only the white witch but also twelve species of sphinx moths. Three of these species are familiar to me, especially one, the widespread species *Manduca sexta,* the subject of my PhD dissertation at UCLA. Merian at the time had no names for the insects; it was not until 1758 that Linnaeus's *Systema Naturae* appeared and laid the groundwork for modern nomenclature. Linnaeus used her illustrations in some of his species descriptions.

When Merian spent two years in Suriname, she had compelling reasons to leave home and take her daughter along. One was apparently to escape her husband, another to join a religious sect at their home in that country, and a third to follow her passion of finding new caterpillars and studying and drawing them. She received financial support as well. I don't know how she succeeded on all accounts, but she discovered many species of then-unknown animals, and she probably raised caterpillars on their respective food plants to the pupal stage and then on through the adult moth and butterfly stages. The food plants of any of the sphinx moths I captured were far beyond my scope. I could not know the species of plants, did not find a single sphinx moth caterpillar, and would not have been able to find out "to see what it would change to." All such detailed knowledge of interrelationships is reserved to those who make their home in a place long enough to get to know it and care about it. Our stay was short, but it seemed long.

Finally, at the end of three weeks we had packed our gear and collections for the museum and were anxiously perched on our folded-up tents at the agreed-on time for our long-anticipated pickup. Our food was gone. Beer had been nonexistent. We joked about drinking cold beer in Paramaribo. It seemed as if we sat there forever — but finally we heard the rhythmic rumble of the chopper in the distance. I don't recall who heard it first, but there it was, finally coming around the edge of a mountain — the sweetest sight we could possibly imagine. Everyone jumped up, to be ready the moment it reached our LZ that we had carefully cleared of brush and where our baggage was placed in a pile to be loaded in a minute or two.

This time the weather was fine and we were in great spirits, serenely humming along over the virgin forest and peering down, and I was again looking for a possible harpy eagle nest. But after an hour or so in the midst of this reverie we heard a sharp loud *bang*.

Before we left Paramaribo three weeks earlier, we had met a Suriname diplomat whose parting words to us were "Good luck — at least statistically you won't *all* get killed." Now I had a sudden doubt about that pronouncement.

Glenn was the first to step out after we landed in Paramaribo and the roar of the engine stopped. He closely inspected the windshield directly in front of his driver's seat. He wiped a spot of dried blood there, and I pulled off a tiny feather that had been caught by a windshield wiper. I stuck this trifle into my notebook. I think it was a pigeon feather. But what I really took with me from this trip was a perhaps trite realization: There is no place like home.

HOME CRASHERS

THE HOME I BUILT FOR MYSELF IN MAINE WAS ONLY A SIMPLE
cabin, but already in the first spring the guests started moving in.
I could hear the activity of deer mice at night, and a woodchuck
dug its den under the floorboards. A pair of phoebes nested on a
log under the roof on one side, and a colony of white-faced hornets
built their nest on another. Red ants found the roof space under the
metal sheeting quite appealing, because they and their larvae got
warmed there by solar heating. Carpenter ants renovated a partially
hollow log (and were later raided and demolished by the red ants).
And a greater welcoming: a pair of flickers excavated their nest hole
into the side of the cabin and fledged a clutch of seven young in late
summer. Less appealing and not at all welcome, though, the cluster
flies moved in that fall by the thousands, ladybird beetles by the
dozens, along with a few stray green lacewings. I saw a mourning
cloak butterfly inspecting the outside of the cabin, and it may have
found a cozy spot as well. In short, my home had become a biological hot spot. Toward many I had few objections; the cabin has space
enough. My cabin was not like the highly desired real estate of the
tight and tidy mud nests of cliff or barn swallows, which house sparrows, *Passer domesticus,* will routinely usurp and then renovate to

their own specifications. Nor were my cohabitants as unpleasant as bedbugs, which hide in the cracks and come out at night to suck your blood.

Opportunistic home crashers can, as shown by many examples of the social insects and especially by ants, in the long term become permanently entrenched residents. Ants' nests are soldier-defended fortresses filled with huge potential food resources, most notably the ants' own eggs, larvae, and pupae. And yet a host of different kinds of insects, collectively called "myrmecophiles" or "ant lovers," often move in. The ant "love" in this case refers mostly to eating in addition to homing, and myrmecophiles have their home crashing honed to a science. They dupe the ants by mimicking their host's appearance, behavior, and scent so that they are treated as though they are one of them and are fed by the ants. Some, and they include true bugs as well as beetles and larval butterflies, are effectively wolves in sheep's clothing, turning the ants' otherwise vigilant home guarding on its head.

It probably took millions of years of evolution for the myrmecophiles to perfect their strategies. However, in the cases where the guests take nothing, or may instead contribute, the process of acceptance could be quick, and indeed it might almost literally be "instant." I'm referring to some birds living in/on our homes, and also to cases of birds nesting peaceably within the lattices of the large stick homes of raptorial birds. Flickers that make their nest hole in a cabin are a rarity, but in North America, robins, house sparrows, house finches, cliff swallows, and European starlings often nest on our homes, and eastern phoebes and chimney swifts now nest almost exclusively with us. In Indonesia, swiftlets, *Aerodramus fuciphagus*, nest in colonies and make the edible (to us) nests for "bird nest soup" entirely out of spittle, used as a glue and a building material, to nest on the ceilings of rare caves. But in the 1990s they started nesting in the upper floors of the new modern

"Western" homes. Kisaran, a town in Sumatra of seventy thousand people, now has three hundred swiftlet "hotels." The swifts' moving from cliffs into our homes led to "farming" swift nests for income, and swiftlet "hotels" were attached to apartment buildings to invite their occupancy. In part the switch is due to scarcity or lack of original nest sites. Chimney swifts, *Chaetura pelagica,* in North America now use chimneys almost exclusively to nest in, as a stand-in for hollow trees. But elimination of nest sites does not explain all of home crashing by birds. The cliffs where eastern phoebes once nested are still here, although phoebes now prefer to nest on or in human dwellings. In Europe, the swift, *Apus apus,* used to nest on still-available cliffs as well, but now they nest almost exclusively in small openings of buildings, which they apparently prefer. For the most part, we encourage these guests to stay with us. And many are willing guests. I made a list of those birds which in Europe and North America habitually to sometimes associate with human structures and/or the nest boxes we provide, and I noted thirty species that are welcome guests. In direct opposite, we reserve loathing for other home crashers, like bedbugs, for example.

Bedbugs belong to the insect order Hemiptera. They are "true bugs," as opposed to the colloquial name for almost any insect. If you know any bedbugs, they probably belong to the species *Cimex lectularius.* It is a mammal-nest dweller that was once a scourge and was then defeated through chemical warfare, but in the past fifty years it has been making a comeback, due to evolving resistance to pesticides, and now lives comfortably in upscale hotels and slum tenements alike. Another species, *Rhodnius prolixus,* perhaps more commonly known as the "kissing bug" because it feeds on peoples' faces, may be one of the most despised insects on Earth. (It was, however, the favorite animal of renowned British physiologist Sir Vincent Wigglesworth, who gained his fame from using the kissing bug as the ideal experimental animal. With it, he laid the foundations of our knowledge of endocrinology, for which the Crown allowed him to add "Sir" to his already distinguished name.)

Lou Sorkin, a forensic entomologist at the American Museum of Natural History, calls bedbugs "the most loathed insect in the United States today." I suspect part of our loathing comes from our lack of defenses against these unwanted houseguests that have perfected their tactics through millions of years of trial and error, all the while staying with us with little chance of leaving. We can at least swat at a mosquito after it lands on us, and we can hear it coming. But what can you do against a bug that crawls silently at you at night out of a crevice, sneaks up on you while you are asleep, and then begins its work by injecting an analgesic so you won't wake up until it is too late? Each blood meal can yield not just a welt, but hundreds of eggs, hundreds of new bugs, and so your home can in short order become bug infested. There are only limited options for a remedy. The most common is to leave home, which is what some birds do when their nests become infested, even if it means leaving their babies behind.

Bedbugs and also fleas can be especially harmful in social situations where they have many hosts reliably available, such as in a dense bird colony. The team of Charles R. and Mary B. Brown from the University of Tulsa have made a detailed study to find out just what it means for colony-nesting cliff swallows to be invaded by bedbugs, specifically one species out of eighty-nine of the family Cimicidae, namely, *Oeciacus vicarius*, known as "swallow bugs." They found that in southwestern Nebraska, several hundred individual swallow bugs can inhabit a single swallow nest. The victimized nestlings respond to their frequent bloodletting by dying or exhibiting a great reduction in growth rate. When the researchers fumigated heavily infested colonies, they found that the surviving offspring at these fumigated nests were markedly increased in size relative to those in unfumigated nests. The adult swallows are apparently aware of the danger these unwanted hosts pose in their nests, because either they avoid reoccupying previously heavily infested nests, or they abandon those that are heavily infested. As I learned from a correspondent, other bedbugs of the same family also inhabit the nests of swifts, though apparently not in great numbers.

This correspondent, Klaus Reinhardt from Sheffield, England, was a stranger to me. He identified himself as an expert on bedbugs. He wrote to me because "while reading your book [*The Snoring Bird*] suddenly the name of one species of bedbug, which I had always thought a little strange, got meaning." The name of the bedbug he referred to was *Paracimex gerdheinrichi*. Nothing more was known about it except that it came from Indonesia. It is now in the collections at Berlin and was described and named in 1940 by the parasitologist Wolfdietrich Eichler. The locality label states: "Celebes, Latimodjong Gebirge, Oeroe, 800 m, Aug. 1930 (G. Heinrich) in nest of *Collocalia spodipygia*." *Collocalia* is a swift.

I knew then that it was named after my late father, Gerd H. Heinrich, who as a young man of thirty-five years left Europe, on March 16, 1930, accompanied by his wife and her sister, and another woman, to spend two years in the wilds of Celebes (now Sulawesi), at the behest of the renowned ornithologists Erwin Stresemann of the Berlin Museum and Leonard Sanford of the American Museum of Natural History in New York, to bring back birds, especially a presumed extinct bird, a rail named *Aramidopsis plateni*. So, why did he send a bedbug to Berlin? And why was it named after him? As with much in biology, it probably has something to do with home.

Bedbugs don't travel around on the bodies of their hosts; instead they hide in their homes. My father must have collected bird nests, specifically to then try to catch the pests. His usual method, which I recall from watching him when I was a child, was to put the animal and/or its home, such as a mouse nest, into a white bag, seal it tight, and then later check to see what crawled out onto the cloth, where it could be picked off. Bedbugs were the least of the catch from most nests. In addition to the swallow bugs and swift bugs, birds' homes are also favorite residences of fleas, mites, and the maggots of some species of flies.

The seriousness of unwelcome insect guests may be hard for those of us who have chemical defenses that we can order up at will to grasp. Those who can't may pay dearly. Once, in Tanzania, I came to an abandoned Maasai *boma* — a circle of mud/cow-dung-plastered huts with dirt floors in a circle surrounded by a perimeter of heaped and woven thornbush to keep out the lions from the central area that holds the cattle at night. It was, I thought, a romantic setting, one where I would have liked to live. So why, I wondered, had all the people left? The question was, I think, answered the moment when, clad in shorts, I entered one of the huts. In the darkness I almost immediately experienced a creepy-crawly sensation on my legs. When I stepped back out into the light, I saw thousands of fleas ascending my bare legs. I stripped naked and fled. That option, which worked for me, does not work for a baby bird confined to its nest, especially if those guests may be almost invisible.

A neighbor, Bob Heiser, had a phoebe nesting on a beam in his garage with five young that were almost fully feathered out, but one day he found all five chicks on the ground. One of them was dead. He returned the live ones to the nest, but the next day they were again on the ground, but by then only one was still alive. He put the live one back into the nest and sat at a safe distance to watch, expecting to see the baby jump out. But instead what he saw was the adult phoebe return to the nest, pick up the live chick, and drop it out of the nest. A week later the phoebe was incubating again, but in a new nest on a light fixture, closely adjacent to the old. Normally phoebes reuse their old nest for the second clutch. What happened here?

I have observed the same nest-emptying phenomenon in "my" phoebes in my shed on a couple of occasions also, and although I didn't sit and wait to see how the babies left the nest, I did know that the cause was mites. In both of my cases the nests were crawling with hundreds, maybe thousands, of dust-particle-size mites. The baby birds had been captive in the nest. The mites are so tiny that it would be impossible to pick off the individual ones, and their numbers were overwhelming in any case. The babies may have either hopped out on their own or been thrown out by the parents in a case of "throwing the baby out with the bath water."

Getting rid of small nest parasites might involve preventive home maintenance measures. No cases are known where birds respond directly by applying miticide to their babies or nests, but some birds have been thought to incorporate aromatic greens into their nests that could discourage nest parasites. But this is at the time when the nest is being built, not afterward in response to mites and/or other parasites as they appear. Nevertheless, home maintenance behavior is indeed well developed in many birds, and it takes some fascinating turns. It starts with nest hygiene.

Nest hygiene starts at nest building. A ground-nesting bird may scrape the ground, clearing it of debris. A woodpecker, nuthatch, or chickadee excavating a nest cavity picks up the loose chips on the bottom of the nest hole, carries them out, and drops them. In one recent comic twist of this behavior that made the news, a car wash owner in Frederick, Maryland, was losing "significant money" each week from one of his machines. He suspected his employees of having a key to the change box and stealing his money, so he installed a camera to catch the thief. Photographs showed a bird sitting on the change slot, with three quarters stacked on top of each other in its bill. There was no apparent explanation, and several people e-mailed to tell me the apparently puzzling story. Like almost all the amazing YouTube postings where the relevant details are left out, this one was not puzzling at all; a pair of starlings had found the loose-change cavity a convenient site in which to build their nest, and they were cleaning out the "trash" to make space to build it.

Nest hygiene is the behavior of removing foreign or harmful material from a nest. Phoebes, for example, carry off the hundreds of fecal droplets that their babies produce in the nest. They usually "catch" them immediately after their baby offers one, an event conveniently programmed to occur when a parent is present, namely, immediately after the baby is fed (or induced by the parent palpating the nestling's cloaca with its bill, as I observed in flickers). I have watched fecal pellets accumulate under phoebe nests only when it is late in the second clutch's development and when there have been two clutches in the same nest in a single season. Such nest hygiene is practiced in most small perching birds, and it probably functions to reduce the disclosure of the nest location to potential predators, as well as to prevent the soiling of feathers. Almost all warblers and other birds with open nests practice such nest hygiene, although those nesting on cliffs or in hollow trees that predators seldom breach practice it less or not at all. Those encased in solid fortresses or safe locations may, instead of packaging their feces in easily grasped and transported packages, projectile-defecate liquid "mutes" instead. That is, they literally "shoot" their feces out and away from the nest.

Nest hygiene functions to maintain the welfare of the young, but that also involves something much more important than cleaning up wastes, namely, distinguishing rightful residents from the unwanted ones that insert their young (in the egg stage) into their home. Since many birds lay their eggs into the nests of other individuals, the parasitized parents must discriminate against foreign eggs — against those that pawn parenthood onto them at the expense of their own reproductive effort. Foreign eggs may (as in cuckoos) result in the outright murder of the whole clutch, or (as in cowbirds and in birds that lay their eggs into the nests of others of their own species) reduce an unsuspecting pair's reproductive output or at the very least increase their workload.

Nest hygiene, in the form of destroying or expelling foreign eggs, has evolved in many birds as a counter-strategy of others' egg dumping. Elaborate strategies of color-coding the eggs have evolved, where the colors and color patterns of parasite eggs match the host's eggs ever more closely, and hosts make ever more differently colored eggs and ever finer discriminations in an ongoing arms race. Mistakenly accepting parasitic eggs can potentially cost the hosts their young through parasitism, whereas too-fine discrimination of differences could result in them ejecting their own offspring.

Whether or not it is appropriate to throw an egg, one that appears to be "different," out of the nest depends on the balance between the likelihood of being parasitized and the cost of accepting a parasite egg. Chickens, for example, readily accept almost any foreign egg because there is little cost in having more babies, since the adults don't have to feed them. Ravens, I found out by experiment, accept almost anything, including red-painted chicken eggs, potatoes, and flashlight batteries. Their tolerance is not based on inability to distinguish these items from their own eggs. Instead, it is because they are unlikely ever to be cuckolded; their large size, high vigilance, and strong territoriality make it easy for them to detect strangers coming near the nest, so the benefit of acquiring nest-content-ejecting behavior is less than the cost.

Consider now the very consistent and potentially high cost of accepting practically invisible parasites that can weaken and kill. Such parasites are no problem when they are few, but those few can become the seeds of destruction if they multiply fast enough, until it is too late to combat them by direct one-to-one encounters. They are the aforementioned fleas, bedbugs, mites, and lice. Discrimination is not always possible; maggots of certain species of flies may either be harmful if they suck the blood of the young, or helpful if they eat their wastes, as in some woodpecker nests.

Problems arise when the home is used for a long duration for each of the deposited parasite eggs to grow to an adult and for those adults in turn to stay and reproduce for multiple generations. That is what the bedbugs, mites, and fleas do, because the larvae and the adults feed on the same thing: the blood of the ever-present living occupants. In the arms race between these unwelcome houseguests and their hosts, it would appear that the guests have the upper hand. Often they do, but being too successful would guarantee their own eventual extinction, because they would eliminate precisely that which they need to survive on. They exist *because* they are held in check, perhaps since their hosts have evolved defenses such as fast growth rates of the young and/or frequent change of homes.

Biochemical immunity and/or biochemical defense derived from medicinal plants are also potential options against the often invisible home crashers. Many insects, such as ladybird beetles and monarch butterflies, have evolved to incorporate toxic compounds from plants into their bodies that make them distasteful to potential predators. We have learned to use some of these compounds as repellents, and it would seem natural that some birds protect their nests with them as well. But, as with most things, in practice it is far from straightforward.

Starlings are one bird species that appears to exploit plant chemicals in nest defense. Recent studies by Helga Gwinner and colleagues in Austria have demonstrated that the European starling, *Sturnus vulgaris*, a cavity-nesting bird, may incorporate green plant material into a nest. Dry material is usually a preferred nesting material for most animals, because it is the best insulation and prevents rot. But since these birds did incorporate green aromatic herbs, it was thought that their potentially negative effects are counterbalanced by medicinal value, specifically as a miticide. Gwinner's studies were aimed to test this hypothesis, and various other possibilities and nuances were evaluated.

Perhaps the first surprise was that, although in starlings females are the main nest makers, it was the males who brought in the greens, and only near the end of nest building when mating occurred. Proximally, the greens functioned as sexual attractants — they were thought to attract the females to the males who provided

A just-hatched broad-winged hawk on a freshly inserted green fern frond. No greens are added during the month-long incubation, but after the young hatch, the greens are brought almost daily until the young fledge.

them. However, as with most sexual attractants, for them to be effective it helps if they have value. It turned out that, although the greens that the males brought did not affect mite populations in the nest, they did affect the health of the young, enhancing their immune function and increasing their growth rate.

I have on occasion looked into starling nests and had dissected one nest into its component parts and counted the number and color of the feathers it contained, to contrast them with those in a nest of tree swallows in an adjacent nest box (the starling had mostly brown feathers, the swallows almost exclusively white), but I found not a single green sprig in it. On the other hand, I found huge fresh green ferns and cedar sprigs lining the nest of the broad-winged hawk. These greens were so big, so prominent, and so fresh that they "knocked my eyes out" in their conspicuousness, yet I knew of no studies that examined the phenomenon of covering the entire lining of a nest with greens.

"My" hawk nest was located about ten meters up in the triple crotch of a sugar maple tree. Broad-winged hawks build a rough nest structure of sticks, and they line the nest mold with chips of dry bark. Finished with this lining, these birds, like most others, do nothing further at the nest after laying their eggs except incubate for about a month. However, in the hawk nest that I climbed up to examine almost every other day, there was something unusual: it was only *after* the young hatched that the nest contained the foot-long fresh green fronds of ferns and cedars. Even stranger, the hawks continued to bring fresh greens about every day or two for the next month until the young fledged. Were these greens some kind of remedy to discourage bugs?

The chronology of the insertion of the greens alone precludes any likely involvement of the various hypotheses previously proposed. Furthermore, in comparison with the amounts of greens incorporated in the starling nests, those in the hawk nest were massive. Since the greens were routinely and consistently replenished only after the young hatched, it had to do with the young. But what? My so-far-untested hypothesis is that the greens serve to provide a clean surface for the meat that the hawks bring in for food.

Hawks routinely bring a surplus of food into the nest for the young and leave it on the nest platform for them eventually to dismember and eat. This behavior would be comparable to a human parent routinely lugging in a hunk of beef and dropping it onto the floor of the home to then let the kids scrabble over it. Broad-winged hawks have their young in the hottest part of the summer, and meat that is not immediately eaten can spoil quickly if infected with bacteria. Uneaten meat scraps and a soiled "floor" seeded with bacteria that had multiplied from previous meals would hasten spoilage. A layer of greens, replaced frequently, would prevent debris and soiled material from accumulating and so retard food spoilage.

We too use herbs, chemical sprays, and swatters to keep the bug crowd as well as bacteria at bay. I didn't do that in my home and as a result inadvertently adopted an unusual and perhaps uncommon houseguest. It was of the eight-legged kind whose specialty is catching the six-legged winged kind. I gladly let her stay, and she entertained me for two years, to then spark a scientific project.

CHARLOTTE II: A HOME WITHIN A HOME

There are spiders living comfortably in my house while the wind howls outside. They aren't bothering anybody. If I were a fly, I'd have second thoughts, but I'm not, so I don't.

— Richard Brautigan, *The Tokyo–Montana Express*

A LOUD BUZZING ALERTED ME, AND LOOKING UP I SAW A large bristly fly caught in a spider web that I hadn't known was there. As the fly continued to buzz, a large orb web weaver dropped from a ceiling beam and pounced on the fly. In seconds the black fly was wrapped in silk and looked like a mummy in a white sack. The spider then attached the fly by a short strand of silk, and touching it with one of her hindmost pair of legs, which she held out stiffly, she then casually walked back to the log on the ceiling from where she had come, turned around, and while holding the fly by its front pair of legs, pulled it to her mouth in what looked like an embrace, and followed up with a very prolonged "kiss" with her fanglike chelicerae. Five hours later the empty hulk of the fly dropped onto my desk. This July 11, 2010, meeting was my first introduction to what would continue to be a housemate for the next year, and through

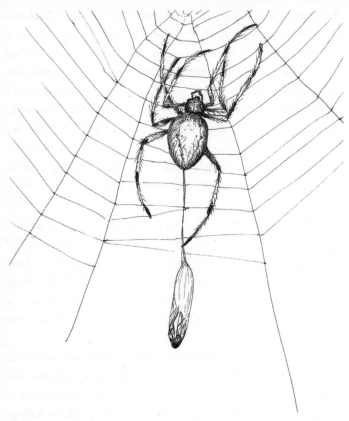

Charlotte, while on the underside of her slanted web, has wrapped a moth in silk and attached it to a short thread, which she uses to carry the prey up to her lair above the net. She uses a hind leg to hold the prey away from her web while carrying it.

the next summer as well. She had made her home in mine, and in honor of the fictitious spider E. B. White made famous, I called her "Charlotte."

E. B. White featured an imaginary orb web spider, possibly *Araneus cavaticus* (which he named Charlotte A. Cavatica, or Charlotte for short), living in a barn in his famous children's novel *Charlotte's Web*. The orb web spiders, genus *Araneus*, spin famously intricate and beautiful silk webs for catching insects. These webs are not just their hunting tool. They are also their territory to which they attach a home, which is usually a crevice, or a hoodlike roof they create by bending a leaf and holding it together with silk strands. The webs consist of sticky "catching threads"; radial "spokes" for holding the sticky threads; "bridge threads" like guy-lines for holding the net up; "signal threads" that inform the spider, through vibrations she feels in her legs, if prey is struggling in the net; "drag lines" for access into the web from her simple home; and still other threads for wrapping prey and eggs and transporting prey. Orb web spiders even employ a temporary silk; while beginning to make their web, they insert spirals of silk into the center of their orb that help to stabilize the spokes while they attach the sticky catching threads. They remove the central stabilizing threads (by eating them) before they finish adding the remainder of the sticky threads.

The silk for all these constructions is stored as liquid, which the spider extrudes according to use from six apertures ("spinnerets") at the tip of her abdomen, and this liquid hardens into the silk thread when it is exposed to air. How a spider keeps all her threads in order to produce an intricate orb web is a miracle beyond my comprehension, especially after every step of the process is enumerated, which makes it only more incomprehensible rather than less. Although E. B. White touted Charlotte, his "extraordinary" spider, for imaginary web-spinning prowess, she lived like many do, in a barn for a summer, and she died in the fall like all northern orb web spiders are reputed to do, namely, at the age of one year, after they have laid their huge clutch of eggs and wrapped it in a silk sac for safekeeping through the winter. (I'll have more to say on orb web spider life span later.)

Adult females make silk egg sacs in the fall and guard them for their remaining short life. In the spring, the dot-size spiderlings exit the sac and extend threads that catch the breeze to waft them aloft until they eventually settle far from their origins. I am familiar with several local species of this family, the Araneidae, and my "Charlotte" was, as I later verified, indeed the one named "barn spider," *Araneus cavaticus,* the one featured by E. B. White. Another of the local large orb web spiders, the pale white and yellow six-spotted orb weavers, *Araniella displicata,* weave leaves together to create an envelope-like shelter, leaving an entrance at the bottom through which they exit to their net below that is strung across a small clearing in the brush. The common garden *Argiope* spiders, *Argiope aurantia,* perch usually directly in their webs on the milkweed and goldenrod in my field. A marbled orb weaver, *Araneus marmoreus,* has for the past two years made its web at the same spot at the front of my outhouse. It uses the roof to hang its web and spends most of its time poised to pounce from a cleft under the roof. Spiders, though, don't always have such discriminating home specifications. In many other kinds of spiders, such as nursery web spiders, Pisauridae, their homes become the place where the eggs hatch and then become the "nursery" of the small young where the adult female stays to guard them. Others, such as wolf spiders, Lycosidae, carry their egg sacs until the eggs hatch, and then continue to carry or guard the young after they hatch. They carry their young on their back in a tight backpack-like ball. Thus, baby carrying is not a strategy of only some Hemiptera bugs, marsupials, and primates, all of which (except for one exception we will examine later) do not make a home.

White's Charlotte was deemed "extraordinary" because she saved her friend, Wilbur the pig, from being turned into bacon by spinning words such as "Some Pig" in her web directly above Wilbur in his pen, to be read by the amazed farmer. I'm more fascinated by the story of live spiders and was wondering what they might have to say by their own behavior, rather than what they thought about pigs. But, as with the famous imaginary spider, the skills that made my Charlotte amazing also related to her web.

There is no way of knowing how long she had been in my home, but I suspect she had lived elsewhere before I first met her, because she was already huge then. By three weeks later, in August, more flies had made the mistake of flying to the screened window, trying to get out, and then instead getting into her net. I suspected she would soon convert her food into eggs rather than growth.

One day, at 7:30 a.m., a little before my breakfast, I decided to feed her first. I ventured out into the warm sunshine and grabbed a bumblebee from among those that are at this season numerous on meadowsweet blooms. Back inside, I threw the bee into Charlotte's web. She almost instantly "dropped" along a thread the seventy-five-centimeter distance from her home, the lair on the ceiling, into the center of her web. The bee remained motionless. Charlotte stopped dead in her tracks. But only for a second. Then she held her front feet to some threads, jiggled the web with them, and almost immediately turned directly to the bee, which was ten centimeters beyond her. She grabbed it and in the same motion began to twirl it with seemingly most of her eight legs, as a skilled juggler on her back might rotate a ball on her feet. Silk spewed from the spinnerets and in ten seconds her prey looked like a white mummy except that it was buzzing and I could see the bee's barely moving legs laboring inside.

The silk wrapping the rotating bee was attached to the net at two ends, like a spindle, and the silk she shot out from her spinnerets was not a single thread like one from a spool. It looked like a triangular white sheet that spread out as it left the hind end and enveloped her victim. It was presumably a spray of liquid that solidified as it was extruded. Her hind legs were not only rotating her prey, but also spreading the spray of silk strands apart to make them into a sheet and to help wrap the prey in a straitjacket. Once the prey was fully wrapped, she released the bee from the net by biting through an end of the spindle, and then dangled the bee package by several centimeters of thread from her rear spinnerets. She then held the bee off the web with one of her stiffly held hind legs and ran up the web using her seven remaining ones to run. The bee package dangled on that one thread the whole way up to her lair (after watching several hundreds of this seemingly delicate maneuver, I saw no instance of a spider dropping its prey), where she attached it, and then turned around and bit into the bee. She held it by her front-most pair of legs and sat almost motionless for the next two hours, presumably injecting enzymes through her chelicerae as through the needle of a hypodermic syringe, to digest and then suck out the bee's contents.

Were these behaviors executed by blindly following a script? What would happen if the sequence were interrupted? What would she do if a second prey entered her web before she had finished the repertoire of handling a previous prey? Which "program" would she follow then? To find out I threw a big bristly (Tachinidae) fly into her net while she was still feeding on the bumblebee. She instantly released the bee (which stayed in place in her lair) to drop down into her web and grab and wrap the fly. But instead of now carrying the prey up into her lair as she normally did, she killed it in situ and left it secured in the web where she had caught it, to then walk back up to her lair and resume feeding on the bee. Would she now forget it? And how about another distraction, a third simultaneous potential prey?

An hour later, while Charlotte was still feeding on the bee, I threw in a white admiral butterfly. It thrashed violently, but Charlotte did not budge, perhaps because the commotion was too vigorous or because she was no longer hungry. To test, I threw in another bumblebee of the same size as the first. It, too, was ignored. So, she was either not hungry or could not triple-task. The butterfly escaped.

Charlotte finally finished her "breakfast bee" at noon and dropped its by-then dry husk. But instead of staying in her lair, as I had expected, since that is what she "always" did after finishing a meal, she casually walked down into the net to retrieve the dead fly. I was surprised because normally she stayed all day in her lair and came down into her web only to capture struggling prey in it. Since orb web spiders are practically blind, she could not have seen or felt the fly, which was past the point of jiggling the net. Since Charlotte had never spontaneously walked around in her net during the daytime, she must have remembered the fly that she had killed and left there earlier. (Wolf spiders, Lycosidae, and jumping spiders, Salticidae, build no webs and hunt by stalking and leaping on prey by vision alone. Jumping spiders have a set of two huge eyes directed forward, plus three more pairs pointing in different directions.)

At 2:00 p.m. I threw still another bumblebee into the net. Charlotte instantly left the fly in her lair to get the bee but hesitated when she was close to this much bigger bee. It struggled for the next two minutes as the spider continued cautiously to palpate the net with her front legs. This bee eventually escaped, and Charlotte returned to her lair to continue feeding on the fly. The net, due to the struggles of the two large prey, was by now badly torn.

The net was not repaired by the next day, August 7. I had read that a spider repairs its net every night. Charlotte had apparently not read the book, or the book is only about the very hungry spider. Is she still interested in capturing prey? I wondered. So — I threw in another bee to find out. Her answer was a resounding yes. She vaulted a meter or more down to capture it, wrap it, give it the bite of death, and immediately take it up to dine on in her lair. And she didn't repair her net the next night, either. Her handling technique involving all eight legs was, as before, not only a theme of fascination for me but entertainment to the kids visiting me at the time.

August 26. Charlotte was, as usual, in her lair. She had made a beautiful new symmetrical web of thirty spokes and sixty concentric catching threads the night before. In quick succession I tossed her two small bumblebees (the most common and easy-to-get insects at that time). She captured and silk-wrapped them one at a time but took only one of them up to her lair. Four hours later, while she was still feeding on it, I tossed a third one into the web. She ran down and tried to wrap it but twirled it only a few times and then left to resume feeding on the first bee up in her lair. This third bee escaped. But the second one from the morning, now five and a half hours later, was still securely in the net and moving now and then because Charlotte had neglected to kill it. Now she finally went down to bite and kill it, then climbed back again to resume her meal. It was a small bee, and she not only sucked out the juices but also chewed it to a pulp. In a few minutes she dropped the head of that first bee, and then left the rest of the carcass briefly, only to return to it and resume feeding on it. (She dropped all the empty silk-wrapped hulls onto the table below, where I was able to see the remains of her meals, both the hard parts that hadn't been ingested, and the round liquid fecal spats of what had passed through her gut, imprinted on paper I had placed for record keeping under her lair.)

The next morning the bees' remains were on my desk, and Charlotte was motionless in her lair. I tossed a small grasshopper into the net. She rushed down into the middle of the net, then stopped as if lost. Unlike the energetic bees, trapped grasshoppers often do not budge. She gradually wandered all over the net, tweaking threads here and there, presumably trying to induce her prey to move to try to escape — a mistake that instead quickly results in their capture. The grasshopper still did not respond, but after three minutes she was finally within a couple of centimeters of it. This time when she tweaked the net, the hopper did move slightly and she was on it in a second. In about five seconds she rolled it in silk and then, with it attached by a thread from a spinneret on the end of her abdomen but held to the side off the net guided by the left hind leg, she headed back to her lair. But one of the hopper's long hind legs was sticking out, and since it had not been held far enough away from the web, it got caught in it. Instead of continuing, she turned around and came back to reattach the hopper to a thread, and then resumed taking it up to her lair. But after ascending a short way, it again got caught in the web. But this time she was several centimeters from "home," and she continued and went all the way in. Yet, a minute or two later she "remembered" that her hopper was missing, because she came down to bring it up.

A big fly was handled differently. It also got caught in the web on the way up, but she continued all the way up to her lair and after a short pause came back down on a guy-string, bit two of the threads holding it, and attached another short one from her rear. Due to the angled web and the low position in the web, she and the fly then swung about ten centimeters away from the web where she dangled loosely, while dangling the fly. She then ran up the length of the one-meter vertically dangling thread to return to her lair, pulling along the huge fly five centimeters behind her.

In late afternoon an approximately seven-centimeter-long dragon-fly came into the cabin through the open door, flew to the window, and landed in Charlotte's web. She ran down toward it immediately, stopped within about fifteen centimeters of it, and then retreated. No further reconnaissance was made as the dragonfly continued in its struggles, and after fifteen minutes it escaped. Was this prey too large? I tested by tossing an ant into a remnant of the now badly torn net, and after only a few seconds' delay Charlotte came down from her lair, found it, and took it. I resumed testing her discriminations and tastes by tossing in a fuzzy caterpillar. She came down to investigate it, but then went back into her lair. Not hungry? I tossed in a bumblebee — it was taken.

She rebuilt her web by the next morning, but with only thirty-six circular orbs, compared to the sixty on the day before.

As the summer rolled on, Charlotte continued to be a constant presence a meter over my head at my desk. I fed her occasionally, and she fed herself on the cluster flies that were starting to come into the cabin in October. Most of the trees had by then shed their leaves, and temperatures in the cabin were at times low enough to form ice. Her routine had seldom varied through summer and fall, but one day — on October 29 — I didn't see her at her usual home made of silk under the ceiling boards. I felt the loss of her and presumed she had, according to the prevailing lore of orb web spiders, died. But on November 11, after I had been away a few days, I found her barely a meter from her old lair. She had come to rest on the underside of a ceiling beam where she was hidden by a loose piece of bark, and there she had molted; a cast "skin" (cuticle or exoskeleton) clung to some silk beside her.

The cast skin was quite important: it meant that she had not yet reached maturity, despite all of the assisted hunting throughout the summer. As far as I could determine, she had not grown much during the summer, and she had been large to begin with. Was she already several years old when she settled by the window? That cast skin indicated that she still had some growing left to do.

May 7, 2011. The winter was long past, and I was back at camp. The weather had lately been depressingly cold and rainy. Insects had just started to fly again. The birds were quiet and the trees leafless, but fiddleheads were starting to poke out of the ground, and to my great delight, but not surprise, Charlotte was still at the *precise* spot on the ceiling beam where I had last seen her in November. She had survived the winter; when gently poked, she moved. I was surprised, because I was aware of the spider lore that orb web weavers grow to adulthood in one season, as in E. B. White's story.

By May 19, 2011, Charlotte had left her hibernation spot and returned to her home, the exact lair in the right-hand corner of the second south window where she had lived the previous summer. She again built a symmetrical web against the window, also in the same place as the summer before. It was still early in the season and I saw neither flies nor any other insects at the window. Might she be hungry?

Bumblebee workers were not yet available, but blowflies were aplenty just outside my door at a chipmunk carcass. I caught one and let it fly into Charlotte's net, and the instant it hit the web and got snagged on one of the sticky threads, Charlotte dropped down faster than I had ever seen her move. Slam, bam, in a second she had the fly and, after a quick bite and rollup in silk, had it dangling below her on a thread and was running back to transport it into her lair. Then she began her slow "kissing" embrace, and two and a half hours later when she was done sucking the juices out of it, she also chewed it to a pulp.

Her appetite was also expressed in other ways; as soon as it got dark, she came out of her lair and took up a position in the center of her web, where she stayed to "hunt" even while continuing to chew on the ever-tinier remains of her blowfly. While so engaged, one after another blackfly flew into the net (I had had the door open during the day). These tiny flies didn't move once trapped, but she kept shaking the net to try to locate them. And after she did, one at a time, she added them to the same food bolus she was already feeding on. Since she was locating nonmoving prey, presumably by causing it to jiggle due to the momentum of the energy she had put into it, I needed to ask her something I had not asked before: Could she distinguish prey from dead non-prey?

One at a time, I threw into her net pieces of bark, carrot, and a seven-centimeter sliver of wood. Charlotte did not respond to any of these immediately, but after a few minutes she did slowly walk down from her lair, jiggle the net, and locate the objects. Leisurely, she cut (bit) the threads holding them and dropped the bark and carrot chips out of the net. (The net was angled about ten degrees from the vertical, and the objects dropped below the net, not onto it.) But the long wood sliver flipped over and got caught again. She recut the strings holding it, and then it did drop. She then went to the center of the net, jiggled it once more, and with nothing left in the net, she returned to her lair on the ceiling.

Charlotte's hunting territory attached to her home is like the forest a wolf knows near its den. The wolf hunts by smell and by sight, paying special attention to movement. The spider hunts by feeling the movements of prey, triangulating, then pinpointing the location of potential live and even dead "prey." She unerringly returns from this hunting territory at any time directly back to her lair, the tiny home where she spends most of her time in the day, but comes out into her web and stays there full-time all night.

May 22, 2011. Charlotte's new web was now of similar size to last year's; today it had twenty-seven spokes and forty-three rings. Not unexpectedly after her seven-month fast, she continued to have a huge appetite. She slammed down to catch a blowfly, leaving a blackfly on which she had been feeding up in her lair. She wrapped the blowfly and took it up and later came down to catch another one before finishing her feed. She wrapped this one, too, but left it in the net to go back up to her previous meal. But later she calmly walked down to retrieve the one she'd left hours earlier. In the evening, she resumed her now-preferred night-hunting position — the center of the net. (In five others of her kind and size that I kept in the cabin two years later, all came down within several minutes of each other at dusk and returned again as synchronously at dawn. Net building occurred only at night.)

June 3, 2011. I gave her a moth, and whereas she had hauled up her most recent prey guided on her right leg, this time she used her left as I'd seen her do many times. She was ambidextrous and flexible, using her left hind leg, right hind leg, and occasionally both legs, and even her mouth, to haul tiny prey.

June 6, 2011. At 8:05 a.m. I threw a dead blowfly into the net. By 10:21 a.m. it was still not taken, even though I jiggled it three times to try to mimic a struggling insect. Was Charlotte not hungry? A minute or two later I threw in a live moth, and she came down immediately and grabbed it, but in the process the dead fly fell out. I threw it back in and got the same non-response from her. It appeared that she did not want this fly. Maybe a different one? So I put a live deer fly into her web, which brought her down to hunt immediately. She jiggled the net with the two flies in it and now grabbed the dead, previously rejected blowfly, apparently making a wrong inference, namely, that the net movements by the live deer fly had come from the location of the dead blowfly that she had rejected. But she wasn't duped totally: she calmly cut the big blowfly out of the net, dropped it, and then located and took the deer fly.

By early July, Charlotte's upkeep of her web declined. She hardly reacted to any insects I offered. I destroyed the whole ratty mess of her web on July 8 to find out what she would do then. During the next night she installed seven meter-long lines down from her lair and attached them to the top of a Coleman lantern that was then on the table below. They appeared to be spokes of a web. Three days later, she still had not added circular concentric strands to her web. I again removed the existing strands and the lantern.

On July 23, the web was rebuilt, but only crudely. Nevertheless, it did catch a big bristly fly, and she processed it. After another three days, she again strung a series of about a dozen threads in a funnel-like formation from the ceiling by the window in a radius of about forty degrees, converging into each other and attached directly to my table. In the middle of them she placed a rough orb, but it was no more than thirty centimeters in diameter, rather than the usual four-times-larger size. It looked more like a sloppy maze than a web. Charlotte seemed to have lost her touch, or was she perhaps overfed and doing only the minimum to get by?

In September, when the first red was already showing on the maples in the swamps and it rained day and night, Charlotte was in her web at night and up in her lair on the beam in the usual place in the daytime. She took a grasshopper. I then tested her tastes (on September 30) by throwing her a stinkbug, wondering how she might react to defensive chemicals. No problem: she rolled and wrapped the bug in silk but then left it in place. It looked like a rejection, but later that evening she came back to it, hauled it up for her supper, and sucked it dry. On October 4, 2011, I offered her a *Polistes* wasp, which has a long flexible abdomen with a reach and a stinger at the end of it. She came right to it but kept it at "leg's length," and her legs were long. After a lot of maneuvering, she did manage to silk-wrap the wasp so that it became unable to flex its abdomen far enough to contact her, after which she approached its front end and leisurely bit into its head. Throughout this process, the wasp's long and flexible abdomen kept trying to reach around to her. After she killed it, she started to transport it up into her lair, but then just kept the wasp where it was and fed from it there in her web, which had again started to look like a ratty mess. Two days later when she refused a syrphid fly, an otherwise never-refused prey, I took down the tangled web, hoping to watch her rebuild it that night. But she didn't work on it.

The next morning, she was not in sight. She had left her web and was on a ceiling beam toward the inside of the cabin, at almost the same spot where she had overwintered before. To get there from her lair by the window she had to cross three ceiling beams. She later moved to an even darker spot and there spun some silk surrounding herself, so she meant to stay. I was afraid she might dry up there because it was near the stove, which I would use all winter. So I installed her in a veggie crisper with some bark to hang on to and put her on a shelf.

When I checked on her three months later, her abdomen was discolored, flaccid, and shrunk. If she died because she had reached the end of natural life, she didn't leave the expected egg case. Maybe I had chosen the wrong overwintering spot for her. With no possibility for intake of food or water, the overwintering home must be cool and moist enough (in the case of hibernators) to conserve energy and water. Survival then may depend on stomach contents. There are secrets here, but only other spiders like Charlotte will tell, if you ask her just right.

All of the spider lore that I read about, and heard from three spider experts, and as per E. B. White's famous story of Charlotte, is that orb web spiders lay eggs in late summer that they ensconce in a fluffy silk cocoon, then die in "late autumn." The eggs overwinter and hatch in the spring, and the almost microscopic spiderlings then "balloon" far and wide in the wind, being carried by a strand of silk they extrude. But, although it is well known that adult orb web spiders die in the fall after reproducing, I found no information about anyone who tracked individuals in a temperate seasonal environment from the time that they hatched until they reproduced, to find out if they accomplished that feat in one year, or two, or ten, or thirty.

My Charlotte left me with mysteries, and one of the main ones was how she could produce offspring, since she never gave signs of wanting to leave home. How could she have found a mate? Being saddled with a web might make it difficult to leave home, because what would she eat? Why didn't she ever try to relocate to a new home, one where she might have found a mate? Males supposedly make no webs, but if a male is found in a web it is presumably because he had entered it to find a female. Because of this information I had been convinced that Charlotte had been a female, even though she had not left an egg case. More or less by chance, answers emerged the next summer from a new spider.

I had grown fond of Charlotte in her own special spot over two summers, and I missed her the next year, 2013, while wrapping up the writing of this book. I was, however, by then keeping a vigilant eye out for orb web spiders, especially at places that were cavelike, where those like Charlotte might make their homes. I ended up keeping many, of various sizes, at my cabin window and in cages. Most were various-size replicas of Charlotte in every way. But one was different.

In early May, when the first leaves started to appear on the trees and the first insects started to fly after a long cool rainy period, I found spider webs under a rock overhang at a road cut that I pass regularly on a daily run from my cabin. And, as hoped, another large spider that looked like Charlotte was perched above the web, clinging to the underside of the rock.

I checked on this spider several more times, and finally on June 9, thinking about some of the mysteries that Charlotte left me with, and also finding a discarded cardboard coffee cup there by the road-side, I stopped to catch the spider in the coffee cup, to bring it back to the cabin. It was time, once again, to have a spider housemate (and later several).

This spider's pair of palps — the organs at the front end of a spider that are analogous to antennae in insects — were enlarged and clublike at the ends, indicating it was a male. But the day after I released him in front of a screened window in my cabin, he had built there a beautiful typical orb web — of about forty centimeters' width. In every way, the web looked like Charlotte's and the other females'. I captured insects, tossed them into his web, and he acted in every way like Charlotte had: capturing, rolling up, and pulling up prey to his lair on the ceiling.

Although I decided this spider was probably a male, I was not certain, because he was huge, and I had read that spider males are small relative to the females, and everyone I talked with informed me that male spiders do not build webs.

The new spider's patterns of gray and black lines and white markings indicated it was an *Araneus cavaticus*, like Charlotte. However, aside from being more leggy than Charlotte, he had in comparison to her a tiny abdomen. So I kept throwing insects into his web. The last one I threw in, on June 18, right after he had made another huge new web, was a drone honeybee, whose body was about the same size as his. He silk-wrapped the drone, sucked it dry, and dropped the silk-wrapped corpse from the lair above his web onto my desk.

After the bee-drone meal, the spider stayed put in his lair, and he ignored all potential prey I threw into his gradually more tattered web. I thought he was sick, but a week later, on June 25, he molted and his already seemingly shrunken abdomen was now smaller still. On the other hand, his legs were a third longer than before the molt. His first leg, the longest, was 4.3 centimeters, giving him a maximum stretch of 9.2 centimeters. The dimensions of the cephalothorax were equal between male and female. He had, it seemed, metamorphosed into the body shape of a runner. He built no new web, and then he disappeared.

This could have been the end of his story, except that the fireworks on the Fourth of July kept me awake, and in my half-sleep I had thoughts that I wanted to write down. I got up, put on a head-lamp, and, holding a pencil and notepad, sat in darkness on the couch downstairs and tried to scribble. There, within a minute or two, I felt a light, brushlike sensation on my bare skin, as though touched with a downy feather. I looked down at my thigh just in time to see a huge spider leave and, without a pause, run along the couch. I recognized the small abdomen and the oversize legs.

And this spider could *run!* In two seconds he had made it to the windowsill, and as soon as he sped along it I started count-ing: "thirty-one, thirty-two, thirty-three" — in three seconds he had traversed the length of the one-meter windowsill. At that speed he could do a thirty-minute mile. I looked at my watch — 12:05 a.m. The spider, on reaching the corner, next raced up the wall and in seconds reached the ceiling. No female had ever acted remotely like this. Apparently male spiders of this species, after they mature, run around at night, and that is why the females don't have to. Instead, they can stay put to continue catching prey, conserving and gath-ering resources to make their masses of eggs as their abdomens continue to expand. Males can then devote the rest of their lives to the mate chase. But if so, they can mate only once or perhaps twice. (Their pair of palps are not tactile organs like the female's — they are instead sperm transfer organs. During mating one or both palps break off inside the female.)

Three days later I found him again, this time running around upstairs. Still, no web had been made, and I wondered if he would feed at all. I captured him and put him in a thirteen-cubic-centime-ter cage to test if he would feed on the flies, bees, beetles, and grass-hoppers that I offered live — he ignoring all (a female *A. cavaticus* I kept later in the same cage routinely captured prey). And, after a month without taking food, he was finally on his last legs and held still enough for me to sketch him.

The evolutionary "strategy" of this spider — where the female is sedentary and feeds and the male is highly mobile and does not feed when adult — is not an altogether unfamiliar one: in some species of moths (such as the aforementioned bagworm moths, the locally common tussock moth, *Orgyia* sp., and some winter-flying Geometridae) females lack wings and have only rudimentary legs, while the males have large wings but lack a digestive tract. They do not move beyond the cocoon out of which they emerge, deriving all of their food from the actively feeding larval stage. In ants and termites, there is (in addition to different body forms related to caste and sex) a serial change of body form and associated behavior instead; both sexes are highly mobile before reproducing but shed their wings as soon as they settle for the rest of their lives into a home to reproduce.

Note: Having learned where spiders like Charlotte live and build their webs, I found twenty-three of them at a deserted camp complex in the woods in mid- to late August. At first I saw only huge ones, like Charlotte and the male had been, but the longer I searched, the smaller the size of these orb web weavers seemed to become. The tiniest individuals that I eventually found were mere specks. They appeared to hang in space, could be seen only if they happened to be in a web, and against a uniform dark background, and at just the right angle in the right light. With a hand lens, I saw that the tiniest of these spiderlings had created almost perfect replicas of adult webs, but these webs were woven of such a fine silk that they were, for all practical purposes, invisible.

Of the twenty-three individuals, nine had an abdominal diameter (as measured by a caliper) of 12 to 13 millimeters, two of 7 to 8 millimeters, three of 4 millimeters, two of 2 millimeters, seven of

1 millimeter or less. In terms of approximate volume, the spiders' nearly round abdomens calculate to about 0.52 cubic milliliters for those of 1-millimeter diameter, to 905 cubic milliliters' volume for the largest, those having 12-millimeter abdominal diameter. These size differences occurred near the end of August, and after the one month more that they had left before going into hibernation in that year, they had grown only slightly or not at all. All the largest ones laid a round packet of hundreds of yellow eggs, to which they clung until they dropped off dead in late October.

Spiders are the most common arthropods I encounter in hibernation in the winter woods, and I believe by a conservative estimate that Charlotte was at least five years old when I got her, and more likely very much older than that. The duration of the orb web spider life cycle has been hugely underestimated. I think that if Charlotte had matured from a hatchling to egg-laying in one Maine summer, she would have been an "extraordinary spider" indeed, one measuring up to E. B. White's original.

THE COMMUNAL HOME

We become human only in the company of other human beings.
— Paul Rogat Loeb

THE NESTS OF MOST SONGBIRDS ARE SMALL AND HIDDEN, made by one bird or a pair, and function only to hold, hide, and protect one clutch of eggs and young. Those of the sociable weavers, *Philetairus socius,* of the Namib and Kalahari deserts of southern Africa, do much more. They are the world's largest and most populated tree houses, or nests, which may weigh several tons and range up to six meters wide and three meters tall. Over a hundred nesting chambers or apartments contained in one of these communal homes are refurbished and reused and new ones added over successive generations, often for over a century; one generation inherits, builds on, and profits from an environment created by another.

In the sociable weaver's native habitat of the Kalahari Desert, temperatures range from minus ten degrees Celsius to over forty-five degrees Celsius in the summer. Like a human apartment building (to which it is impossible not to draw comparisons), this weaverbird's home is a year-round community's living quarters that shelters all from direct sunshine and also protects them from rain, drought, and cold. Scorching summer temperatures demand water for evaporative cooling in many animals, and most birds regularly need water to get rid of excess body heat by gular fluttering, a process similar to panting where increased air movement in the moist throat area accelerates evaporative water loss. But by retreating into the communal nest at high temperatures in the summer, the sociable weavers save water otherwise needed for heat dissipation, enough so they can get by without drinking.

Not needing to be near water hugely affects the possible range where the weavers can live. Conversely, at low air temperatures in winter, when the birds spend nights in the nest, they save the energy otherwise needed for shivering to keep warm, thereby saving food. Some apartments used by pairs for rearing their young in winter become dormitories for up to five adults, and when there is frost outside and they huddle together, the inside temperature stays fifteen degrees Celsius above that outside. Thus the sociable weavers' communal nest is a valuable resource other than just for breeding; it is also a year-round dormitory and a refuge. Perhaps understandably, the young of these already sociable birds are reluctant to leave their home after they fledge, and they stay, "earning their keep" by helping their parents feed subsequent broods.

Unlike most weaverbirds, the sociable weaverbirds don't "weave." Their nests look like huts complete with a thatched sloping roof made from grass stems that sheds rain. The nest grows through the years as the birds insert dry grasses into the bottoms and sides, as new apartments are added. Each of the separate compartments of over a hundred breeding pairs is lined with soft downy plant material and has a separate entrance about twenty-five centimeters long and three centimeters wide, built of downward-pointing spiky straws that help as snake excluders. As a result of the close proximity of nest entrances, the underside of the communal nest has a honeycomb appearance.

Communal nest of the sociable weaver. Entrances to individual compartments are on the nest underside (right).

Perhaps not surprisingly, what works well for the home-makers also works for some home crashers. The weavers' communal home is also the almost exclusive nesting place of the pygmy falcon, *Polihierax semitorquatus,* and a favorite nesting place for red-headed finches, *Amadina erythrocephela,* and a parakeet, the rosy-faced lovebirds, *Agapornis roseicollis.*

The weavers' nest-building drive, and their attachment to the communal home, is so strong that they not only live in and around their nest all year but also continue their home improvements throughout the year. At the San Diego Zoo, home of the only colony of these birds in the United States, the busy birds are daily provided with dry grasses for their year-round home-making.

· · ·

When confronting phenomena as highly derived as a weaverbird's communal nest, the honeybee's language and social organization, or a skyscraper, we wonder how they evolved. These things may seem incomprehensible when the intermediate steps from "there" to "here" are obliterated, but possibilities can be visualized by comparisons with the well-known that exists. Weaverbirds' nests vary widely, and an examination of their different kinds provides examples for a progression of possible stages from solitary to social behavior sometimes associated with social nesting. The progression from the original to the highly derived may not be entirely correct, but it provides scenarios that can be tested.

To begin with, weavers in general tend to be sociable. Many nest in colonies, and their nests are often in close proximity to one another. Nests of the white-billed buffalo weaver, *Bubalornis albirostris*, of Africa south of the Sahara, often touch each other. One bulky stick nest can become a convenient location to which the next one is attached, until several effectively become a communal nest with separate entrances. This arrangement is also found in another African weaver, the chestnut weaver, *Ploceus rubiginosus*, which nests in enormous colonies, and in other arid scrubland weavers of Africa, such as *Plocepasser mahali* and *Pseudonigrita arnaudi*, both of which tend to form clusters of about a dozen nests and are more socially tolerant than most other weavers.

Parrots provide a comparative example, converging on the same pattern as weavers but from a very different direction. Unlike most weaverbirds, which create their own cavities by weaving bag nests, parrots nest in the already-made cavities of tree holes. But there is an exception: the monk or Quaker parakeet, *Myiopsitta monachus*. This species is a true exception because it is the only one of the 350 parrot species not limited to nesting in cavities. Instead, it builds stick nests in trees, and not just little ones made by mated pairs but giant car-size communal homes.

Since all other parrots nest in cavities, we can infer that the monk parakeets' communal tree nesting is a derived condition that has evolved from ancestors that were cavity nesters. The example then begs the questions of how their tree nest building evolved and why it has not evolved in any other parrot for which choosing cavities to nest in was highly advantageous and so became deeply ingrained in the birds' bag of survival tricks. A radical change to a totally new mode, which would be akin to trying to rebuild a propeller plane into a jet, is unlikely. So how was the conversion from cavity nesting to building a stick nest made, and especially why was it again a communal nest, for specifically this species?

The monk parakeet's ancestral nesting places were likely similar to those of the closely related cliff parakeet, *M. luchsi,* which nests in rock crevices rather than in tree holes. Cliff sites are partially enclosed by walls, and initially cliff nesters may have filled in some of the open space with perhaps a twig or two, and then maybe more. A cliff crack would have provided space for an adjacent nest, or two, or more. "Extra" space would have reduced competition at a nest site. Instead, a nearby nest might have helped to make a wide or long crack more like a cavity, so that a potential competitor was instead a "helper," thus selecting for increased social tolerance. Increasing social tolerance in turn would permit the use of ever-wider crevices that could be substituted for cavities. Nesting next to or in between several other nests then became the monk parakeet's *substitute* for nesting in partially enclosed spaces such as those on cliffs.

Nesting next to each other was a win-win condition for all, and at this (current) point in the parakeet's evolution, hundreds of nests may be crowded next to each other to create a very large communal stick nest where each pair of birds has its separate "apartment" with a separate entrance. Like the homes of humans and those of the sociable weaver, the parakeets' nests are used as a site for other birds to build their homes in. They include the spot-winged falconet, *Spiziapteryx circumcincta,* and two species of ducks, the Brazilian teal, *Amazonetta brasiliensis,* and the speckled teal, *Anas flavirostris.*

The rare socially nesting birds out of the many remind us of three very conspicuous exceptional home-makers among the mammals, two from within the rodents and one from the primate family. In these three, as in both the weavers and the parrots, the home is not just for rearing young. Are they all exceptions? That is, could they have been predicted? If so, then only if we understand the causes. Richard D. Alexander, a zoologist at the University of Michigan with a special interest in the evolution of social behavior, did just that. In the mid-seventies he published and lectured widely on the topic of social behavior, trying to decipher how gregariousness may have evolved to living in a large group with overlapping generations and division of labor, and with only one or two reproducing adults while the rest were all sterile. This suite of four conditions is labeled "eusociality" and it had been found in what were then called "social" bees, wasps, and termites but had not been observed in a vertebrate animal. In trying to illustrate how eusociality (as opposed to mere social behavior) had evolved, Alexander hypothesized a fictitious mammal that, if it were to exist, would be eusocial given certain conditions. His mythical beast was modeled after termites, the eusocial insects that had presumably evolved from ancient solitary cockroaches that had lived in and eaten rotting wood.

Unknown at the time to Alexander, Jennifer Jarvis, a student then working on her PhD at the University of Nairobi in Kenya, was starting to provide the groundwork for his prediction. She was interested in deciphering the biology of a curious type of rodent living in those arid regions in southern Africa where a number of plants have evolved large tubers that are the underground food- and water-storage organs that tide these plants over during long and unpredictable droughts. Her rodent, appropriately called the naked mole rat, family Bathyergidae, relies primarily on these tubers for its food and water and lives "permanently" underground in highly branched tunnel systems that have sleeping and toilet compartments and may extend to over three kilometers in total length. This was precisely what Alexander had predicted for a eusocial mammal, and when that was recognized, these animals' social biology became the focus of intense study by these and other researchers, working both in the field and with captive animals in the laboratory housed in simulated naked mole rat homes of interconnected cavities. The picture that emerged was that the naked mole rats were indeed eusocial; as many as a hundred individuals of one species, *Heterocephalus glaber,* live in one tunnel system, but only one female out of the whole colony reproduces, and she is mated by only one to three males. All the rest of the occupants of any one colony help defend it, plug entrances, and work as excavators extending and making new tunnels.

Other species of mole rats are not eusocial, so again the question was, Why did this one species become eusocial? It probably has to do with the same conditions that produced termites from cockroaches.

Termites' ancestors probably fed on decaying wood, which also enclosed them in a confining space but protected them from predators. Similarly, a naked subterranean mole rat feeding on a huge tuber can make its home there, and the family can remain in relative safety without starving and without venturing outside. A singly excavating mole rat is unlikely to find one of the far-dispersed tubers, and equally unlikely to survive aboveground. But with many excavators in a colony, the odds of one finding a new tuber, after one is eaten, are increased and the costs of sharing small, so social tolerance becomes highly adaptive. Tubers may weigh up to fifty kilograms and supply enough sustenance to last a whole colony of fifty to one hundred individuals for over two months. By their cooperation, the individuals gain by reducing their risk of starvation. The cost is that, in the close quarters, there is opportunity for overpopulation, so neutering becomes the price of food and safety.

Numerous studies after Alexander's insights, and the empirical work by Jarvis's and Alexander's students and others spanning decades, closed the similarities between termites and naked mole rats. The mythical animal proved to be real, validating the idea that the evolution of eusociality probably resulted from confinement of overlapping generations in a safe home.

The adaptations of those mole rats that specialized for underground life include, beyond their eusociality, their almost total loss of hair except for whiskers used as sensory organs, stiff fringes on their feet that aid in digging, and the loss of external ears, loss of vision, and shortening of limbs. All of these apparent "degenerations" are adaptations for underground life. They would be a huge liability aboveground, which then creates ever-greater advantages to staying within the underground colony, creating a spiral that, once started, potentially locked them into eusocial life.

The enclosed space of the home that confines the individuals facilitates the establishment of dominance by the parents over their offspring of succeeding generations, which stay and are physiologically neutered by the dominance hierarchy where reproduction is restricted to the select individuals at the top of the heap. In *H. glaber*, the only reproducing female and her one to three mates have four or five litters per year, of four to fifteen pups per litter. I think the reason why the sociable weavers and monk parakeets have not become eusocial is simple: they still live in their own "apartments" in the communal house and are free to fly around; they are not under continuous surveillance and control by any other specific individuals. But why did other burrowing rodents living in arid regions in other parts of the world not also share the same home and become eusocial? Here the likely answer is food supply.

Animals come together and stay together in a common home for food and safety. If either resource is missing, dispersal is necessary, and domination becomes unlikely. In the American Southwest, for example, there are no plants in the desert that have huge tubers such as those of African plants, so desert rodents such as kangaroo rats, pocket mice, and ground squirrels, although they have underground homes in burrows and plug up the entrances of their homes with earth during the day, must venture out at night to collect their food — seeds on the ground surface. They cannot stay safely sheltered almost permanently in a home as the ancient cockroaches did in excavating wood and using it as their food and their home at the same time, or as the naked mole rats do while confined in their underground tunnel system. But although they cannot avoid being on the ground surface, they can reduce their vulnerable time there. And, unlike the naked mole rats, they have evolved cheek pouches to carry home many seeds to then eat in the safety of their underground homes.

Primates are, with one exception, one of the least creative animals in home building. Baboons sometimes use caves for overnighting. Chimps pull a few branches together in a treetop to create a crude platform for a night's sleep. Surely monkeys and apes were not limited by intelligence in evolving to build nests or homes, if they needed them. But among all the primates, only humans build homes. Our extreme exception is so obvious that we are almost forced to consider what it is about us that tipped the balance to make us so very different from our close remaining relatives. As before, we must begin with basic biology.

Primates, for the most part, have only one young at a time, and they carry their baby with them rather than leaving it in a shelter. Those species of crustaceans, insects, spiders, and mammals (most famously marsupials) that do carry their young also lack any kind of home-making. Given the convergence of such unrelated groups, one can assume independent origins for lack of home construction, arising from a common denominator. So, then, how come *we* diverged from the mold to evolve from tree-climbing apelike hominids with little or no need for a nest to build, to living in skyscrapers? How do *we* differ from other primates, to end up with a strong predisposition both to become highly social and to build shelters? For that we need to search for fundamental differences from the other primates.

In contrast to all other primates, our young are at birth altricial in the extreme, and they continue to depend on constant social support for at least five or six years, and on training for still more years. There is debate about why we ended up with babies who are born naked and helpless except to accept food and express wastes like those of many birds. What that biology means, though, is that, like birds and other mammals with helpless (and especially multiple) young that could not easily be carried around, we probably originated from a unique predecessor that required a "nest," or we built a safe "nest" and adapted to safe homes, which permitted our young to be left in a safe place so that they could *become* altricial. As in birds and as in other mammals, the home and the altricial became linked and now are part of the same package. Furthermore, with us it was not just the babies who were bound to that nest. The mother was bound to stay in its vicinity as well — at least more so than in any extant ape where the group can and does travel at will.

Presently existing ape females can be more easily mobile with a baby than naked pre-humans would have been, because ape mothers have fur and a relatively horizontal back for a baby to ride on when they are moving around, and also because ape babies are born with both an insulating covering and the dexterity to hang on to their mothers. In contrast, bipedality created a huge constraint in free movement while carrying offspring — especially because the mother may have had more than one offspring at a time. Furthermore, meat-eating hunter-gathering pre-human hominids needed more, not less, mobility than apes, while having less means to carry even just one young and keep it comfortable at the same time. As a consequence, a pre-*Homo* equivalent mother (before the advent of tools used to carry babies) who was tied down to a home meant that, as in species of monogamous birds, she needed a helper on whom she could depend to be at least an occasional provider. True, Bushmen carried their babies in slings, and Native Americans used a rack on the back where the babies were as secure as a quadripedal baboon's are riding on their mother's back and clinging to her fur, but that was unlikely to have been the case *until* we were human.

The upshot of all this is that with the male pre-human leaving mom and a helpless infant or two in order to go find food, it was imperative that the offspring reside in a safe place with a guardian, or two, or three, or more. Likely more: we may not have resided in hollow logs or been subterranean, and it is a cliché to say so, but a truth nevertheless that some of our ancestors, and already those more than a million years ago when our predecessor was likely a different species, found a home in caves.

Caves may have been ideal homes, because attack there could come only from one side, but the rarity of caves, and the necessity that they be accessible to good hunting territory, must have generated competition. That is, as with termites and naked mole rats, a good home is a strong inducement to stay at home. The offspring, as those in beaver lodges when the parents produce more young, may have been reluctant to leave and "face the world." They may have to have been evicted instead. One thing is sure: if they were allowed to stay long enough that there was overlap of generations and crowding, they would have had to compromise some authority and priority, and/or they would have had to provide something of value in order to be tolerated as co-inhabitants. They could have been helpful in defense, in the hunt for food, and in child rearing. Social tolerance, division of labor, cooperation, and subservience to rules or individuals, of a dominance hierarchy in the group, then became necessary and hence valued survival traits.

Human home-making could then have evolved in a manner not much different from that in weaverbirds, and the making of homes was a huge step that allowed us to expand into all sorts of previously unavailable habitat. First it was in Africa itself, and eventually it reached far beyond. Portable skin houses allowed the Native Americans to live on the plains and hunt the bison there, and similar family houses brought other peoples out onto the wide-open steppes of Asia. Sod houses enabled humans to survive the cold winters in northern Europe and Asia, and the invention of homes made from the snow and ice permitted the invasion of the Arctic, with access to a huge bounty of prey along the shores of the Arctic seas. But it was the innovation of multiple-occupancy houses — apartment complexes — in conjunction with agriculture that made the extreme human density of our cities possible. With those, we have also expanded vertically.

We have no way of knowing what our original homes looked like because they were almost certainly perishable and would not have fossilized. The structures of extant human homes in many environments likely are close to what might have been. Given our imagination, we would have been aware of what other animals did. Perhaps our homes at first converged to resemble (and possibly mimic?) those of the resident birds'. The Pygmies of the Ituri Forest in central Africa build round huts next to each other. Woven from branches and covered with leaves, these huts are reminiscent of weaverbird nests. The Maasai, and the Pueblo Indians, built and still build ovenlike adobe huts with a side entrance that in form resemble the homes of swallows, which are often colonial. Maasai huts are arranged in a small colony serving as a semipermanent home that is surrounded by a wall of thornbushes as a defense against lions. The Anasazi of the American Southwest built cliff dwellings with clay and straw, building materials much like those used by swallows, phoebes, and some other birds. At the "Cliff House" at Mesa Verde National Park, some 150 rooms are packed adjacent to each other. In Europe, earlier people probably lived in natural cavities in rocky cliffs, much like the ancestors of the monk parakeets. These people, like the monk parakeet ancestors, presumably learned to plug the holes of the cliff openings with wooden material. This would have been a first step to building a shelter with a wall or two against a cliff, and ultimately of attaching other homes to it, to create more walls, until more and more rooms were created, so that a communal structure with numerous entrances to separate identical family rooms allowed for efficient climate control, a settled life, nurseries for the young, and division of labor. However, all traces of similar homes of predecessors in Europe, Africa, Asia, America, or Australia would have vanished much as any bird nests of grass, leaves, and sticks would have vanished. What we see in caves represents a sample that is likely hugely biased by differences in preservation.

What then were the implications of many pre-humans or early human individuals sharing the same home? First, several individuals living together permits specialization, a phenomenon that is almost always linked with eusociality but can exist independently of it. Specialization concerning division of labor can result in a massive increase in efficiency, which then can lead to tapping into new resources, and therefore result in growth and permit the existence of even more individuals.

Eusocial bees are an excellent example of this. There are worldwide hundreds of species of bees that are solitary. A female makes her modest home, such as a burrow, and raises her own offspring there. Some species, as the bee specialist Charles Michener described, may have nests with side burrows and/or cells where several young are reared at once or overlapping with each other. Most solitary bees, under selective pressure to provide for their offspring, specialize in foraging from the flowers of specific species of plants (or from a group of species with similar flowers). They thus must restrict their seasonal activity to those often-narrow seasonal times when their food sources are in bloom.

Eusocial bees, on the other hand, tend to be generalists that learn skills of flower handling, so different individuals from the same home often specialize on different flowers. Having multiple specialists in the same home where there is sharing widens the bees' niche. Bumblebees are a good illustration of this. Bumblebees are fairly general flower foragers; they are not hard-wired specialists for any one flower kind. By following any one individual bee from the time she emerges from her cocoon in the nest until she dies, one sees the following: the young bee stays in the nest the first few days and does any number of various house chores, but when she leaves the nest to become a forager, she flies almost indiscriminately to sample various kinds of flowers. Eventually she settles on one main one. She will soon know where to find these, and she will fly without hesitation to them and learn to handle them faster, to extract their nectar and/or their pollen. She may have, aside from her "major," a "minor" flower, which gives her flexibility in case her major stops blooming. Thus, one *Bombus fervidus* worker from any given home may major on jewelweed and minor on New England aster, a second one may major on aster and minor on jewelweed, a third major on mint and minor on a yellow composite, a fourth major on a lily and minor on wild carrot, and a fifth perhaps collect honeydew from aphids. Together they have all the food-producing sources covered, both at any one time as well as through their oft-shifting availability from early May until late fall. Not only is this constant availability of food resources possible because of their social organization, it also makes their social organization possible in a self-reinforcing cycle. As in the eusocial naked mole rats, the risks of running out of food are greatly reduced.

Multitasking vastly expands the bees' niche, but then an inevitable consequence of that capacity of hugely enhanced efficiency in tapping resources sets the stage, close on its heels, for another consequence, namely, following the ability to tap vastly more resources, maybe even seemingly inexhaustible ones, they will with certainty *become* limiting. This is due to the fact of the biological "law" of reproduction proceeding until it encounters limits. If there were no limits placed on reproduction and resources were inexhaustible, any planet where this (impossibility) occurred would eventually explode. That is, given some reprieve from epidemic disease and predators, there would come a time when new selective pressures would emerge as a means of regulating reproduction. There is no animal society on this or on any other planet in which this would not be true. Since we do not see anywhere anything *other than* a stabilization of the populations, it can be inferred that there has been and is very strong selective pressure against catastrophic reproduction.

In all of the social animals, the first and probably the most powerful force that maintains the social fabric is forced birth control resulting from conflict. A termite mound may contain millions of potentially reproducing individuals, but it *actually* contains only one male and one female who reproduce. All the others are sterilized. Each ant, honeybee, and bumblebee colony contains, for the most part, also only one female who reproduces (although several males may have mated with her). Numerically, most of the other colony members are also female, but they remain chemically sterilized as long as the reproducing female is present. Only when the egg-laying female wanes in her reproductive power can and do some of the others develop and lay eggs. If a second reproductive female should emerge while the first is still reproductively active, the younger female will be killed if she does not leave first. One can appreciate the logical necessity of this, because if all of the tens of thousands of females started to lay eggs, it would be impossible to accumulate any resources, and the home with millions would very quickly collapse in chaos. If the bees, or some other hypothetical animal, should by some miracle be able to tap into a new, unforeseen source of food to fuel their economy, they would experience a few halcyon days of growth, to delay the day of reckoning a little further into the future, but to produce in the process a vastly larger debacle.

These considerations concern reality, such as the power of gravity, the speed of light, and the nature of the atom, from which all sorts of inferences can be drawn about our world. Naturally, one has to wonder, since we know concretely how these natural laws apply to termites, bees, ants, and naked mole rats, how might they apply to us who now live in virtual safety from all predators and most diseases?

Given life in a good place, and no conscious effort to limit family size, a *Homo* couple could easily conceive and raise a dozen offspring. Traditionally in agricultural societies, where the couple had more offspring than they could retain, the eldest son inherited the home and the land, the daughters were married off, and the "spare" sons were forced to leave home to seek their fortunes and find a home elsewhere. Population control happened inadvertently through famine, war, and disease. However, we also applied some draconian solutions to control births, in perhaps novel but not necessarily fair or adequate ways. Mostly we instigated various mechanisms to limit sex. The medieval European custom of putting a lock on women's genitals with a metal chastity belt was in parts of Africa substituted by clitoridectomy and/or stitching the vagina shut. In addition, we erect taboos against sex, which in some societies were enforced by threat of death. But the cultural methods focusing on controlling women's wombs run up against resistance, and rightly so, and one has to wonder if there are also biologically evolved, less draconian mechanisms of reducing births under locally crowded conditions.

Is it possible that in some pre-agricultural bands of humans or pre-humans conflict was reduced and the society preserved by some of its members willingly forgoing reproduction? Is it only a wild speculation to imagine a nonfertile segment of the society who, rather than destructively procreating, are creatively aiding? Such neutering is the standard practice that has evolved in the eusocial insects, where as mentioned earlier some members forgo both sex and reproduction and devote themselves to the service and upkeep of the society. This also happens in naked mole rats, where most of the members are and remain sterile by practicing no sex, and they perform vital functions that benefit them and the rest. The Catholic Church, for instance, imposes the neutering option via sex prohibition culturally in its clergy, ostensibly in order to commit them totally to the institution instead of to a family. The church encounters difficulties in this approach, because it does not take human biology fully into account: the forgoing of sex is difficult in a primate for whom sex serves various functions aside from procreation, such as bonding. However, even sex elimination has been accomplished in some tight human societies, such as by castration in the court of a powerful monarch (which closely mimics the situation in social insects).

One could imagine a more benign solution, such as a tight-knit society where some of the individuals in it are sexually attracted to others of the same gender. They would be spared the arduous demands of child rearing and would instead be able to specialize, as bees do, on tasks that require a high degree of attention, learning, and expertise and that would then benefit the whole group. It could be adaptive, and, if so, there is reason to suspect that the frequency of it would depend on stimuli resulting from environmental pressures. I do not know if this scenario is, was, or will be an adaptation in any animal. It can be. But, if so, it would not be in birds, in which sex is perfunctory. It would be in a species that is highly sexed and in which sex is related less to procreation than to recreation. It would be a species with small social groups that compete against one another not by physical force but by intellectual know-how. Such an adaptation is a theoretical possibility, and the bonobo and humans might be possible candidates.

Our genome is highly sensitive to the environment to which it is exposed. As we know from migratory locusts (to give but one of thousands of examples) and many other insects, birds, and mammals, environmental stimuli affect the timing and the kinds of hormones released into and circulating in the blood. Hormones affect behavior, physical development, and other physiological functions, because they control the genetic code and its output. Environment affects the fetus and the developmental trajectory after birth in any of a number of ways that become linked to promote survival. If groups compete against one another, and humans have probably been doing that as far back as we have existed, one group where resources can be used for innovation rather than procreation may do better than another where surplus reproduction brings it to the brink of exhaustion and subsistence living. Birth control for humans — in some cases in the form of chemicals, methods that are another totally new thing on the face of the Earth — in the long term may work to counter the otherwise inevitable collapse.

PART III

Homing Implications

The tie that binds in homing is attachment to place. Yet not all animals are place-bound; some do not attach themselves to any place at all. Instead, they are innately and inflexibly bound to massive numbers of others of their own species. They seek out, orient, or "home" to others that, when en masse, become to them the main relevant feature of their environment. For them, the crowd is, wherever it may be, a place of refuge, as much as a forest is to a whitetail deer, or a field is to a meadowlark.

To many pelagic fish species inhabiting the vast featureless space of the open ocean there is little else to orient to; they have no choice. They swim in huge schools. That be-

havior of using each other as a reference, of homing "to the herd," has evolved as a survival strategy in many animals. It doesn't always work, though, because every strategy in one evolves into a counter-strategy in another; baleen whales capitalize on fishes' herding instinct to catch them. In response to danger, the fish react by herding ever closer to each other, and in the presence of one of their greatest predators, a whale, which blows a circular bubble net around them from beneath, the bubbles then herd the fish, and they soon know nothing about the whale and only herd ever closer to each other. With its gigantic maw equipped with pleated plates, the whale then ingests the whole herd in a gulp. Had they each stayed separate, the whale would not have been a threat.

THE IN AND OUT OF
BOUNDARIES

Good fences make good neighbors.
— Robert Frost, "Mending Wall"

BOUNDARIES ARE NECESSARY FOR ANY LIFE. AT THE CEL-
lular level these are membranes. They separate the chaos of the
external environment so that the intricate structure and biochem-
istry within can be built and maintained. The almost unimag-
inably complex and elegant chemistries of energy metabolism,
genetics, and reproduction could not have come into existence,
or be long maintained, without boundaries that selectively admit
and exclude matter. And like any entity evolved to maintain itself, a home is also a bounded area. This truism does not, however, preclude the equal necessity for "leakage" to allow for new
circumstances as a compromise and an investment for the future.

Nowhere did the opposing functions of home boundaries — to
keep the life-threatening out and the life-promoting in — seem
more obvious to me than in the story of four American chestnut
trees, *Castanea dentata,* that I purchased as seedlings and planted
in the spring of 1982. They are descendants of magnificent ances-

tors that, with a diameter of three meters, grew to be over thirty meters tall. Billions of these trees once graced the North American forest from Maine to Mississippi, and their nuts fed the passenger pigeons, turkeys, bears, and whitetail deer.

Chestnut seeds fall to the ground and can't spread thousands of kilometers by the wind as those of dandelions or poplar trees do, and I saw my four seedlings as a potential baby step toward bringing chestnuts back "home" to a tiny corner of their former range. I planted them in the woods next to my cabin, where I hoped they might be spared from contact with the Asian chestnut blight fungus that was first noted in the Bronx Zoo in 1905. It had been imported from perhaps one or several Asian chestnuts or chestnut trees (which are resistant to the fungus). There was no cellular barrier to the fungus released from Asia in the American tree counterparts, and it spread like wildfire through the forests all across North America. But there were now no more infected chestnut trees in western Maine from which these trees could be infected, so perhaps the seedlings would be protected from infection simply by the boundary of physical isolation. In addition, they were advertised to be from stock that had some immunity to the fungus as well.

The seedlings grew well after I had cleared the competing trees and brush from around them. They had light, and against the usual great odds facing all seedlings in any forest, they did survive the first dangers faced by most young trees — the browsing by hares, porcupines, deer, and moose.

About twenty years later, I noticed something grand: near the end of July the topmost branches of the then-over-six-meter-tall trees were resplendent with white blooms. The trees' flowers have a funky carrion-like smell and attract swarms of flies and beetles. Nevertheless, I did not expect pollination (and hence viable seed), because genetic barriers to inbreeding between these four individuals from one source seemed likely, and there was no possibility for

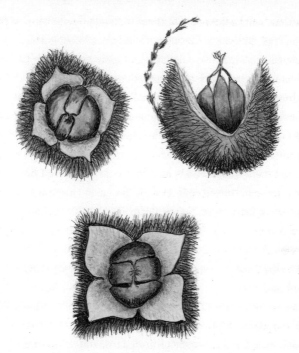

The fruit of the American chestnut tree, right after opening to release the three nuts it contains.

them to receive foreign pollen. Cross-pollination in plants is like sex in animals. It is one of those necessary immediate inefficiencies that scramble the genes to create variety. Perhaps the original American chestnuts were locally efficient, but they had forgone the diversity that could otherwise have saved them; they were uniformly susceptible to the Asian fungus. However, in the fall my trees had fruit hanging from their branches.

An American chestnut fruit is a ball (often called a "burr") about five centimeters in diameter and resembles a hedgehog because it is covered in a thicket of sharp spines. As I had expected, none of these fruits that were soon scattered all over the ground the first few years that the trees bloomed contained nuts (i.e., seeds). Instead, like empty candy bar wrappers, each burr held only the shells of three putative nuts, indicating that they had not been pollinated.

The four trees themselves now faced problems as well, though not immediately from the fungus. Porcupines had practically delimbed and girdled two of the trees within literally an inch (of intact bark) of their lives. I nailed metal flashing around their trunks to serve as a boundary that would prevent the porcupines from climbing and reaching the limbs. Luckily, the badly damaged trees still sent out healthy new shoots, and in a year or two they were recovering by growing new limbs. The trees now, at the age of thirty-two years, are growing faster than ever, and the largest has a height of 55 feet (17 meters) and a breast-high of 53 inches (135 centimeters). They flower every summer and bear fruits with seeds in late October.

As I contemplated the metal exclusion device for defense of the trees from porcupines, it was clear to me that the tree had invented its own mechanical means for protection, but mostly for its delicious, nutritious nuts. Its large seeds are far more prized as food by animals than its bark, and they are enclosed by a wall of spears. A credible boundary to the delicious nuts is necessary and must be relatively, but not absolutely, secure, not only because through evolution *every* strategy is likely to be met by a counter-strategy, but also because this tree species needs animals to disperse its seeds (nuts). But if so, the dispersing animals must be able to eat them, too, or else they would not bother to pick them up to hide them to eat later. According to the old saying "The acorn doesn't fall far from the tree," acorns and chestnuts do fall close to the trees, but they generally can't grow up there. Soon after a young tree starts to grow under its parent, it will become starved for space and light. We now know that part of the solution is birds. Henry David Thoreau (*Journal,* vol. XIV, 1906) was one of the first naturalists to understand this; he wrote, "I have often wondered how red cedars could have sprung up in some pastures which I knew to be miles from the nearest fruit-bearing cedar, but it now occurs to me that these and barberries, etc. may be planted by the crows and other birds." Blue jays and squirrels presumably plant acorns and beechnuts. But I did not know if both, one, or either of them also plants chestnuts.

To see how the tree protects its seeds from being eaten, yet manages to get them planted, required watching.

It's mid-October 2010, and the sugar maples still carry golden foliage while the red maples and ash have shed their leaves, but the chestnut trees still have bright green leaves and hundreds of green fruits hanging from their branches. Within days, the fruits, each containing three nuts packed side by side, will start to open and be released. I knocked down several low-hanging fruits and smashed them open with a hammer, surprised to find plump (as opposed to thin, empty-hulled) seeds. A total of 400 fruits yielded 920 filled seeds (out of a possible 1,200). The flowers had apparently been pollinated after all. Many more fruits were under the trees on the ground. These all had empty seeds, the expected spontaneous abortions of the unpollinated female flowers.

After two days of a strong north wind that whipped the branches about, the rest of the fruits were still not dislodged from the trees, but only four days later, on October 18th, most of the chestnut fruits that were on the trees were finally opening. The four petal-like flanges of each fruit curled their velvety inside surfaces outward, offering their previously guarded nuts. (I do not know how they accomplish this movement of opening; I experimented to test several hypotheses but got no answer.) Blue jays were hopping from branch to branch, searching for the opening fruit and grabbing exposed nuts. They flew far into the distance, presumably to eat or cache, but if they did cache them it was too far away for me to see them do it. Corvid birds routinely cache food, and blue jays are no exception. Ravens dig a hole with their bills, insert the food, then cover it up by using the bill to scrape nearby soil or snow over it, and/or they pick up nearby debris such as leaves and place it over the spot. I had once seen scrub jays (in Oregon) do almost the same thing, but instead of digging a hole, they jackhammered the peanuts into the soil before covering them, much as ravens do. There would be no way for me to know if these blue jays had cached seeds, unless they did not retrieve them all and some of them sprouted trees.

The next year, 2011, hosted a huge masting event of the beech trees (close relatives of these chestnuts) in the New England forests. There had been no beech fruiting for sixteen to twenty years (based on aging the many hundreds of baby beeches), but there was also a record acorn crop. During the first week of October before the chestnuts were ripe, there seemed to be a "highway" of blue jays flying up and down my hill toward and from the beech grove by my cabin. Viewing from a tall spruce tree, I could see how far the jays were flying long after they left carrying the beechnuts, but I could not determine their destination, only the direction. Although jays in the trees picked beechnuts (each fruit contains two triangular seeds), they ignored the nearby still-ripening chestnuts. But even after the chestnuts were ripened, the jays still ignored them, possibly because the beechnuts were more abundant, easier to get, or maybe easier to crack open to eat.

The blue jays I saw on the chestnut trees never tried to hack open a chestnut fruit, and I wondered if the nuts/seeds themselves, which have a hard leathery coat, may be impenetrable to them. When the next year a blue jay started coming to my bird feeder filled with black sunflower seeds, I had an opportunity to find out. I placed fifty chestnuts in the feeder among the sunflower seeds (which the jay had been hauling off for days). The jay immediately took the chestnuts in preference to the sunflower seeds, and in twenty successive visits to the feeder it took them all. Three times the bird flew with a single nut up onto a branch, held it to the branch with its feet, hammered it open in about a minute, and ate the contents. Clearly, this jay preferred chestnuts to sunflower seeds, and it had no problem opening them. However, when offered many at once, it appeared to try its best to haul them off quickly, presumably to cache.

I added more nuts and continued to try to find out how many a jay takes when it flies off into the distance for presumed caching trips. On average, this jay took three nuts per trip. It flew either low into the nearby woods with nuts, or high up into the tiptop of a tree at the edge of the clearing to pause there briefly, and then launch into the distance over the forest. It flew usually several successive trips in the same direction, and then switched, and for another several nut-hauling trips it left in another direction. But why would it fly so far to cache nuts? Does that give it the time to consolidate memory for a specific location? Do longer flights to and from a cache site make the location more memorable than a bunch of close-by scatter hoards? If so, the long flights must be "worth it," and larger seeds would have a better chance to occupy new territory and also have more resources in them to get a head start in growth.

In 2012, when I was on site full-time, I looked closer. This year, in contrast to before, squirrels were much in evidence, and although the spiny chestnut fruits were immune to penetration by jays, they were routinely breached by squirrels, though probably not to the benefit of the trees' reproduction.

As in the previous two years, all four chestnut trees were laden with green fruits in early October. As before, soon after those fruits started to form, the ground under the trees was strewn with hundreds of aborted infertile duds. Red squirrels, *Tamiasciurus hudsonicus,* ignored them. Instead, they climbed into the trees and scrambled methodically from branch to branch, to snip off green unopened fruits that all dropped straight to the ground. These, however, all contained nearly ripe seeds.

A squirrel usually started at the tree each day at dawn and snipped off three to five fruits per minute, and then continued to work until it had dropped a hundred or more. It then left the tree and for the rest of the day perched on the ground or on a stump and eventually gathered the fruit up. One by one it chewed them open and ate the nuts. Additionally, it sometimes took a fruit in its mouth and clumsily carried it three or so meters to a perch to feed on it. Gray squirrels, *Sciurus carolinensis,* came also, but they never once snipped the fruits off to let them drop. Instead, they fed exclusively while remaining perched on the tree.

Together the two kinds of squirrels removed most of the hundreds of green fruits from the trees before the fruit had a chance to open, hence before any jay could have gathered a seed. I found no whole fruit that had been carried off into the surrounding woods, but the ground under and around the trees was littered with piles of chewed and empty burrs. Jays may have planted chestnuts in previous years, but the squirrels seemed poor candidates. The question was: Did either ever act as nut planters?

As mentioned already, I had seen the jays fly off with chestnut seeds in 2010 (and they may have been doing it in previous years), but whether or not they were helping the chestnut trees to reproduce is a separate question. Squirrels have an amazing ability to locate seeds by scent, and turkeys and mice that abound in the woods are also likely able and eager to eat any just-deposited nuts they might find. Any nut that does manage to make it to the seedling stage would probably be a small percentage of the total number of seeds that were dispersed, unless the seed dispersers do something that helps seed survival beyond what would happen from simply being dropped onto the ground. Those simply dropped onto the ground are likely to be more easily found and eaten as well as more vulnerable to drying out or freezing and not germinating. Possibly deep burial, and/or several seeds in the same place, may help to produce a seedling, because the seeds would be protected from both drying out and freezing and because a predator that goes by scent may overlook a seed after it has presumed to have mined the reward at the end of the scent "trail" that attracted it.

To find out if my chestnut seeds were indeed dispersed and had survived as well, I searched for seedlings, first in the immediate vicinity of the trees where nuts would have fallen or could have been scattered, and then at ever-greater distances. If a young chestnut tree was found in these woods, there could be no doubt about its origin, as well as the distance dispersed. American chestnuts were a convenient species for examining how far animal-dispersed seeds may be planted away from the parent's home ground, because unlike almost all of the other local tree species, the seedlings retain their leaves into late October, after most other trees. This makes it easy to locate baby chestnut trees in the fall because their large saw-toothed, still green or yellow leaves show up brilliantly from afar against the then-prevailing brown carpet of the forest floor.

I found only two seedlings under or adjacent to the four chestnut trees, and so I did not have much hope of finding any in the woods. But I searched nevertheless. To my surprise, I soon found seedlings everywhere, and several of them were nearly a kilometer from the nearest chestnut tree. I do not know the total number of offspring these chestnuts had by then (in fall of 2013) growing in this forest, but I located 158 "plantings" of trees spread out over an area of about 200 acres (81 hectares). One hundred twenty of the plantings grew only single surviving trees. The rest were of two to twenty seedlings (or small trees) in a tight bunch in the same spot. The surviving seedlings and trees are of course the minimum numbers that were actually planted.

Who were the planters? If the nuts had been planted by red

squirrels, they should have been close to the tree, since the squirrels are highly territorial and would not hide their food in others'
domains. They should also have been most commonly in groups of
three, since these squirrels, when they did leave with chestnuts, always carried a whole fruit, which has up to three seeds in it. Unlike
chipmunks, tree squirrels do not have cheek pouches for carrying
multiple seeds. Blue jays routinely carried several chestnut seeds at
once, and they flew far out of my sight with them. Since I found up
to five at one spot at a great distance from the parent tree, one can
infer that blue jays had been the responsible tree planters. Only the
jays could account for plantings of more than three seeds at one
spot. Chipmunks were also a possibility for nearby planted trees.

I wanted next to find out possible requirements of seed survival
such as those that the caching by animals might provide. I had
found that all seeds put on wet peat moss in a plastic bag in a refrig-
erator were dead the next spring. This time I put fifty into wet peat
moss again in a plastic crisper which I left outside on the woodpile.
Again, all fifty seeds were mushy and dead when I examined them
right after snow melt. Fifty that I put (in ten piles of five) onto the
ground (in maple woods) were missing by spring. Of one hundred
buried two centimeters deep, seventy-six were missing, but of fifty
buried ten centimeters only six were missing. I concluded that
burial in the soil is necessary both to facilitate seed survival due
to physical conditions and to avoid predation. Apparently the jays
were doing several things right to bring the seeds to a good place
where they could make a start.

Three seedlings that had started in the woods within 130 to 300
meters of their parents have already reached heights of three to five
meters. They have reached sunlight and are shooting up at seventy
centimeters per year. They are now well on their way to becoming
trees, and they could not have come from anyplace else but the spe-

cific ones I had planted. The chestnuts may not have *fallen* far from the tree, but they had escaped the boundary of their parents. They had left "home" and the otherwise inevitable parent-offspring conflict for light, water, space, and soil nutrients. They were of variable ages, suggesting that American chestnuts do not, like the beeches and some oaks, have their October seed-bearing pulses separated by years without flowering and seeding, which serve as a time boundary that can help maintain the seed predators' populations. Instead, they have tough, spiny fruit that serves the same purpose by being a physical boundary to easy access and predator overpopulation. They exclude birds until their seeds are ripe and ready for planting and may then be dispersed, provided enough seeds are available all at once to satiate the then-available takers.

The chestnut trees' *dependable* annual seed crops, the timing of which would have been staggered over a latitudinal gradient, may have provided a reliable food resource to highly mobile birds, such as the now-extinct passenger pigeons. We will never know to what extent this formerly reliable food helped the pigeons to achieve their enormous population size, but the large numbers of them needed in any one colony in order to be stimulated to breed, as we'll see a little later, would ultimately become the final cause of their extinction.

Thanks to the science and ethical commitment to species regeneration of the American Chestnut Foundation (which breeds blight-resistant trees), the chestnuts are coming back, just as the fish and game departments have returned the deer, the turkeys, and the moose. In my forest, where the chestnuts now grow, the turkeys have also returned. Both are multiplying. The forest is largely a mixture of American ash, yellow and white birches, red and sugar maples, American beeches, black cherries, and some white pines, red spruces, and balsam firs. By planting the chestnuts, I breached an artificial boundary. Species are being readmitted to their ances-

tral home where they were part of a complex ecosystem. By planting four trees, I am helping the jays, and more, make it whole again.

Postscript

As this book was going to press, the *New York Times* asked me
to write a short essay for its op-ed page. The *Times* had no specific
topic to suggest, and I had none on hand, until I thought of the
above-described Hail Mary shot-in-the-dark experience of planting
four American chestnut seedlings thirty-two years ago. The off-the-
cuff abstract, which appeared on December 21, 2013, generated a
strong response from readers.

A main concern of readers was the source of my chestnuts. This
had been a concern of mine even before writing this chapter. I had,
among my innumerable crates and boxes of saved paper, by sheer
luck found where I had written down that I had purchased the
seedlings from the "Westford County Soil Conservation District.
Cadillac Mountain, Michigan." But I could not find this on Google,
and so assumed it either no longer existed or had changed its name
in the intervening decades. The American Chestnut Foundation
was founded in 1983, the year after I had planted the trees from
Michigan, and it began breeding for blight resistance in chestnuts
in 1989. The ACF was unable to help me.

The op-ed readers reminded me, though, that knowing the
source of my original stock is important, because some didn't
believe I had planted real American chestnuts, maybe because it
sounded too good to be true. So, under a new stimulus, I made another search. Eventually I located an American Chestnut Council in
Cadillac Mountain, Michigan, where a man named Tom Williams
offered "American chestnut seedlings" for sale. I called the number
(231-775-7681) and got the U.S. Department of Agriculture, and a
long recorded message referred me to extension 3, which turned

out to be none other than — the Westford County Soil Conservation at Cadillac Mountain! I had read on Wikipedia that the American Chestnut Council had located "a blight-free grove of American chestnut trees of approximately 0.33 acres (0.13 ha)."

"The" map of the original range of the American chestnut shows the chestnut tree only in the southeastern part of Michigan near Ann Arbor, not 190 miles northwest at Cadillac Mountain by Lake Michigan. That map also shows numerous other seemingly relict isolated populations. Most of these populations became extinct when the blight swept over the country. However, could some have survived because they were resistant to it, or were they saved simply because of their isolation?

It was the day before Christmas — and the very last day to include edits in my book. But this one was important, so I left a message on the answering machine at the USDA number I had called, and got an email response from Max Yancho, who identified himself as "Wexford Conservation District forester" of the "Wexford and Missaukee Conservation Districts." I had been misled for years in my search for the origin of my chestnuts by misreading an x for an s! Solving this riddle was the best Christmas present I could have imagined. But it got better.

Yancho wrote: "I received your message this morning about your chestnut trees. I had a hard time hearing all of your message and I hope that I can reach you via this email, but the chestnuts which you planted are very likely from the American Chestnut Council based here in Cadillac. The ACC sends chestnuts all over the United States, and uses natural seed stock collected from native groves here in Michigan. I was able to read your op-ed piece online, and I hope this puts your mind at ease. It sounds like you have been very successful in re-introducing the chestnut back into the Maine forest."

OF TREES, ROCKS, A BEAR,
AND A HOME

The mountain pushed us off her knees.
And now her lap is full of trees.

— Robert Frost, "The Birthplace"

MAKING A HOME IS "TAKING ROOT" IN A PLACE. IT IS GROW-
ing enough permanence to encompass life spans and to leave
physical traces, such as those in the arrangements of rocks or the
planting of trees. In pioneer days in the American Midwest, in
order to claim land, settlers had to plant an apple orchard, which
proved their seriousness about making a home there. This was
the setting for the American legend of Jonathan Chapman, now
known as "Johnny Appleseed." A New England native, Chapman
traveled nearly sixty-five hundred kilometers throughout the Midwest planting and tending nurseries of apple trees. He often went
barefoot, was known for his kindliness and sensitivity to animals,
and, unlike many other white settlers, liked and got along with the
Native Americans.

Chapman was born in Leominster, Massachusetts, in 1774. The
young Johnny would have grown up on or near farms being built

by clearing the land of trees, gathering the rocks into walls, and planting apple trees. Most of those farms cleared from the northern New England woods were abandoned in 1816 after "The Year Without Summer," following the April 5–15, 1815, eruptions of the Tambora volcano in Indonesia. Until the dust settled, it caused violent weather the world over. Frosts killed New England crops in May of that year and produced snowstorms in June. A bitter famine winter of 1817 followed, and people left the rocky New England slopes to settle in the American Midwest, then called the "Northwest Territory." The Krakatoa eruption of August 26–27, 1883 (also in Indonesia), was the second-largest eruption in historic times and caused a temporary drop in global temperatures that helped effect a second wave of local land abandonment and westward migration.

There is now an expanse of unbroken (though often heavily lumbered) forest in northern New England where there used to be fields and pastures. The place names in western Maine in Perkins township are Chandler Hill, Kinney's Head, Gammon Ridge, Gleason Mountain, Potter Hill, Wilder Hill, Lakin Hill, Holt Hill, Hedgehog Hill, my own York (or Adams) Hill, Houghton Ledges, Parlin Brook, Bowley Brook. All except Hedgehog (no hedgehogs exist in America; the name refers to the local porcupine) are named after those who made their homes here long ago. The names of the towns are more fanciful, possibly reflecting less the homes of origin than wished-for Shangri-Las of the original settlers. Within a drive of an hour or two, one can visit Naples, Moscow, Hanover, Berlin, Poland, Norway, Denmark, Sweden, Paris, Stockholm, Mexico, China, and Peru.

Few icons of the past evoke romantic nostalgia like the old stone walls and stone-lined cellar holes that one now finds scattered about deep in the woods. Sixty years ago, when my family settled on an old farm in Maine called "The Old Dennyson Place," it was not far from becoming one of those many memories. There was then not a farm without at least a remnant of "the old apple orchard," where on crisp autumn days men and boys flushed grouse and saw porcupines, the fresh tracks of deer, and often a pile or two of bear dung, which looks like applesauce made from whole apples, seeds and all. Sometimes deep in the woods we would see trunks of fallen apple trees moldering into the ground, usually near some rock-lined cellar hole from which trees grew. These favorite places of boys and other wild animals are almost gone now, and you must look closely to see clues of the former homesteads. But rocks and trees still provide reminders of past homes.

The original settlers and their dreams are past, but there is permanence in stone. Next to my cabin is the crumbling cellar hole of the original York and Adams family farmhouse. Most of the rocks of their farmhouse foundations had collapsed into a pitlike depression where a hefty white birch tree, an American ash, a red spruce, and a maple sugar tree had grown. I cleared the trees and brush and chopped, ripped, and tore out a network of roots before I could start resetting rocks. Like a raven in springtime coming back to a ledge and placing a stick or two where it might build a nest if it didn't slide off, I was at that time playing.

During the clearing and digging out, among the submerged granite blocks that had been part of the foundations for two barns I found rusted remains of hay scythes, horseshoes, ax heads, hubs of carriage wheels, pottery shards with pink and blue floral designs, railings from a horse-drawn carriage, plow blades, odd metal rings, lengths of chain, innumerable square nails, hinges, a blue enameled tin cup, and charcoal. Generations had come and gone and the records of their lives have blurred. But there were stories here, and I thought, "If only trees could talk." Silent for a long time, an apple tree eventually did "talk."

The apple tree that had first caught my eye around 1980 because of its large circumference was almost dead. It had a short but hugely thick trunk with two tree-size broken stubs that branched off about one meter from the ground. Most of the tree had broken apart and had long rotted into the ground, but one of the broken trunks still had a thin live sucker shooting straight up in a losing fight to reach sunlight through a rapidly closing canopy of young ash, maple, and white pine trees. I didn't give this tree much thought then while thinning out a future sugar bush. However, thirty years later, when the tree's one remaining live shoot had died, I wondered how long the tree had lived. Henry Braun, a poet from a kilometer or two on the other side of this hill farm, had written, "It isn't far in Maine to the end of the past." Just how far is it? I wondered. Thinking that the old tree might give clues, I chain-sawed the remains down to take a section of still-solid wood and count its growth rings. After sanding and polishing the rich brown-red wood, I counted on average twenty-five growth rings per inch in the outer wood, although the growth deeper in before the tree's decline had been more rapid. There I counted thirteen rings per inch. I calculated that the tree must have started to grow near 1790. Johnny Appleseed was around twenty years old then. George Washington had just taken office as our first president. It was almost a decade old when John Adams, our second president and one of the recognized most influential founders of the United States, was finishing his term. It was a vigorous young tree when his son, John Quincy Adams, was the sixth president.

The decade of the 1790s when this apple tree made its start was significant for this region in Maine. In a detailed history of the settlement of Weld (*Early Settlers of Weld*), E. J. Foster wrote of this area as a basin that is "formed by the surrounding mountains" and that was "little known except to hunters until the year 1782 when Dummer [Sewall] and his brother Henry Sewall set out from Bath, Maine, in March [when they could travel on the crust of winter snow] to survey the country from the Kennebec river to the Connecticut. In their exploration [the Sewalls and the party traveling with them] crossed this valley [and] — pronounced the land to be good quality and worthy of cultivation. They discovered a pond about six miles long, near which they found several traps; and on a tree was cut the name 'Thos. Webb.' This name they therefore gave to the pond, and also the river, which was its outlet into the Androscoggin."

Webb Lake is about five kilometers from where I was excavating the aforementioned cellar hole, which is seventy-five meters from the remains of the old apple tree.

The eighty-five-square-kilometer area by this "pond," after being surveyed, was later called "No. 5" or "the settlement." In 1816 it was incorporated as the 214th town of the state of Maine and named Weld, which now has a store, post office, and most recently (2010) a coffee bar with Internet access. As a teenager, I used to go to the square dances at the Weld town hall, where Rod Linnell was the caller every Friday night in the summer when I worked peeling potatoes and washing dishes at the boys' camp, Camp Kawanhee, at the edge of Webb Lake (i.e., "pond"). The village of Weld is an easy jog down the steep hill from the remains of the old apple tree and my home camp and clearing.

Nathaniel Kittredge, the first acknowledged settler of the Weld area, arrived in the spring of 1799 to "fell trees, and burn and clear a few acres," and to erect a log house. The following year he brought his family, and in that same year Caleb Holt, the second settler, came and followed the same pattern. Walking in on the snow crust, he arrived in March and "planted the first [apple] orchard in town, and made the first cider in the fall of 1829." These historic dates, as I will show later, are relevant to the history of the old apple tree, and ultimately that of the homestead where it grew.

Cutting the wood from the tree and counting the growth rings had revealed the tree's time in the human historical context. The ecological context of where it had grown up was clear from a glance: the two dead stubs a meter from the ground, the remains of two huge laterally spreading limbs. This was proof positive that the tree started to grow on cleared land! But now comes the kicker: how could there have been cleared land up here on this steep hill over two centuries ago, presumably a decade before the first settler, Nathaniel Kittredge, arrived and "burned and cleared a few acres"? Ten years is well within a margin of error concerning the tree's origins, except that the hill where the apple tree grew would not have been settled until much later than 1799, when Kittredge arrived.

The European pioneers who came after this, to seek better land after the good, more accessible ground had been taken, built their cabins and grew their crops in the bottomland along the Androscoggin River valley, and along tributary streams, and then they came to the shores of lakes. They needed land where the soil was rich and deep and where they had access to drinking water, fish for fertilizer and food, and waterways for travel. The tops of these steep, thickly forested hills with their glacially scarred ledges and repositories of glacial till of jumbled rocks required backbreaking labor of men and oxen to clear before cattle could graze and crops could be planted. There was no water on this hill where the apple tree had grown, unless the settlers dug a very deep well. The huge stone walls near it are but one reminder of the hard work of site preparation required to live up here. Some individual rocks in the impressive stone wall weigh tens to hundreds of kilograms, and perhaps tons.

Asa Adams was likely the original settler of this hill farm; there is no record of anyone having lived here before. His daughter Flora was born on the property in 1858, after the Adamses had occupied it since around 1830. So the old apple tree should have been a mature tree of forty years then. Flora married James Kendall York, who came from a nearby farm; she stayed on the farm where she was born and together the couple had nine children (seven surviving). The hill then became known as York rather than Adams Hill.

When I purchased most of the property in 1977, it was uninhabited and had been left idle for nearly a half-century, although some of the Adams and York families had continued to come up in the summers until at least 1929 to pasture their cattle. Around 1930 the house and two barns of the then-York/Adams farm burned. The fields then started to be reclaimed by forest, and twenty-five years later Phil Potter, my Maine woods mentor, brought me here when I was a teenager. We hunted grouse and deer in the abandoned overgrowing fields and apple orchards that Flora and James Kendall York had planted to raise the Ben Davis variety of apples. These were "good keepers" that could be packed into barrels with straw and carted down the hill by horse and wagon to the railroad station by the town of Wilton. From there they were transported to Boston and put on a sailing ship to England. Generations of the York family have told and still tell how old Kendall was proud of his apples and would graft up to five different varieties onto one tree.

The birth date of around 1790 that I determined for the old apple tree by the old York homestead was, as I have mentioned, extremely puzzling, because this was about a half-century before the estimated time their *orchard* was planted, and because the tree's hefty trunk with two thick branches reaching laterally was proof that there was *cleared ground* here over two centuries ago. But how could there have been cleared land forty years before 1830, when the first settlers — who would have settled the rocky hills last — reputedly came up to the Weld area?

The question nagged, and then I took note of another peculiarity of this tree: it was adjacent to a large pile of rocks right where two stone walls intersected. The remnants of the farm's orchard trees, on the other hand, were all at least forty years younger and grew on what had been ground cleared of rocks, and not one next

Various artifacts that became exposed after digging in the soil near the homestead of the Adams/York family

to a stone wall. Only this tree grew precisely where four large stone walls converged, next to a space that was left on one side of the crossed walls where a team of oxen would have been able to pull their sledge through. The spot where the tree grew therefore looked as though it marked a central and likely well-traveled place in a clearing. But where might the seed or seedling of this grand tree have come from?

The new settlers came inland from the coast. They spread out into the nearby hills and made their homes there. Vincent York, who wrote a historical account of this area titled *The Sandy River and Its Valley*, quotes historian William Allen's formula in his *History of Norridgewock: Comprising Memorials of the Aboriginal In-*

habitants (1846) of how the local settlers made their homes there: "[First] year, cut down trees on five or six acres, and burn ground over in preparation for planting; second year, after planting is done, build log home, cut more trees, move family in before the harvest; third year, build small barn, increase stock; fourth year, raise English hay, rye, wheat, and corn and begin living more comfortably; fifth year, clear more land, increase flocks and herds; sixth year, start pulling stumps, and preparing land for the plough; seventh year, build self framed house if you can." He further noted, "A man was famous according as he had lifted up axes upon the thick trees."

The pioneers who planted apple trees soon after they settled near Webb Lake a few kilometers below the Adams site had come up along the Androscoggin River from the Atlantic coast, and they would have cleared land in an upward and outward direction. They routinely used fire as well as axes to make their fields and farms. There were likely forest fires; there is charcoal under the superficial layers of soil in many places on York Hill, a hint of how the clearing was created. I doubt that Johnny Appleseed came up here, though. But bears were plentiful, and they would have raided apple trees then, as they routinely do now. They would have spread apple seeds in their scat and would have been inadvertent apple tree planters.

Bears, wolves, and ravens were probably so ubiquitous then that few people would have bothered to mention them, although one bear encounter is recorded in Foster's early historical account: Abel Fisk, while coming to the settlement in the autumn of 1808, "got lost in the bog at Alder brook" (a fifteen-minute walk downhill from the farm site and my camp), where he also "lost" one of the two horses pulling his wagon. A few days later Benjamin Houghton (for whom Houghton Ledges, a ridge nearby, is now named) "encountered a white-faced bear feeding on the flesh of the horse." There are still bears here and they routinely drop apple seeds.

If the old tree originated from one of those bear-planted seeds on burned land, then, when the lower valleys became settled and Asa Adams and other people came up into the hills with their oxen in the 1830s, they could have found a mature apple tree. They would have been surprised and delighted to see it as a "sign" of home, a good omen, and therefore an inspiration to plant their orchard nearby. Then, when they were starting to clear the land, using "cattle" (oxen) to drag countless tons of rocks on their sledge, they may have gravitated toward that tree. It offered shade in the summer and fruit in September. Is that why the now-massive stone walls radiate from that spot in four directions?

The tree may have told me as much as it could, and despite the questions about the history of this homestead, I thought no more about it because there seemed to be no way to get answers. But then, by a strange coincidence, it reemerged into consideration.

On a tip from Anne Agan, who has deep connections to the York clan, I talked in 2011 with (the now-deceased) Dr. Albert Sawyer, then a ninety-year-old grandson of James Kendall York and Flora (Adams) York. Sawyer, a retired chemistry professor from the University of New Hampshire, was living in Durham, New Hampshire. He remembered being on York Hill as a child. His hobby was genealogical research of the York family, and he had traced the clan back to various royal personages in Europe, including Charlemagne.

The long and complex genealogy, as such, didn't greatly surprise or interest me. It was remote. But then I was invited (and went) to the annual York Family Reunion in nearby Wilton, Maine, where the old gentleman showed me some dreamy, grainy photographs of York Hill dating back a century. They had been taken by his mother, Helen York. Now I was excited.

Helen, one of the seven surviving children of James Kendall York and Flora Ella (Adams), was born on York Hill and for a while (1916–17) taught in the nearby one-room schoolhouse at the foot of it. At the time there was, close to the schoolhouse, a small community of workers at a small sawmill, Hildreth's Mill. It had been set up to make lumber of the tall pines harvested upstream along Alder Brook. The stone foundation of a dam built to use the water to generate power to run the saws still remains. The lake created by the dam is gone, but the brook is back.

Helen had owned one of the first Kodak box cameras (issued circa 1910), and one of her several pictures, labeled "Kendall York Home in the Plantation with Mount Blue in the Background," shows three persons herding at least twenty cattle and a white horse on a bald overgrazed hill in front of a farmstead. I would hardly have recognized the site as York Hill, were it not for the unmistakable shape of Mount Blue in the background. The artifacts and the stone foundations where I have dug previously pinpointed the locations of the buildings, so I knew almost the precise spot from where the picture was taken. Another picture from about 1911–15 shows Helen's parents standing stiffly with the clearly recognizable profile of the Gleason Mountain–Kinney's Head Hill in the background, in front of the then-new apple orchard. But the third picture yielded a total surprise to me, one that nobody else would likely have been in a position to care about, or appreciate. This picture is of Helen herself, and it was simply and inauspiciously labeled by Albert Sawyer: "Helen York: Reading on a Rock Wall at Home." For me it was not about Helen as such, or the book she was reading, the long white dress she was wearing, or her attractive face and hair. It was all about the tree behind her, and what she was sitting on.

The tree in the picture was an apple tree, and given the date of the picture, it was an unusually large one. But mostly what caught my eye were its two thick and low lateral branches. They looked familiar — they seemed exactly placed to match the distinctive form of that huge apple tree I had known while it was still barely hanging on. Could this be the same tree? But I was indulging in wishful thinking; there was no way I could prove my conjecture. Instead, I tried to visualize the century-old scene when the picture was taken.

I "see" a Fourth of July when the York clan, as they had every year in those years, met for their traditional picnic on "The Ledges" — an overlook a few hundred meters above the house site that afforded a grand view down to Webb Lake and Tumbledown and Jackson mountains. Helen had dressed up in her "Sunday best" in anticipation of this event, and she and her four sisters had bantered with her mother, Flora, as they cooked the Sunday meal. The day before they had done the washing, which they hung on a long line (that is seen faintly in the background in Helen's other photograph of the farm itself). After the cooking, and after the women had dressed up, the sun was coming up over the ridgeline of Gleason Mountain and shining on their two apple orchards and the pasture and two hay fields. It was a morning for celebrating, and Helen admired the farm buildings — the house with attached shed and two barns next to it — and saw her father, Kendall, and two of her brothers in front of the house with the cattle and the white horse that pulled their carriage when they traveled down to the village of Weld by the lake. Mount Blue showed tall and clear to the north that day. Helen got out her camera and took pictures, and her mother suggested that she take one of her daughter as well. Helen was hesitant; she didn't like to be photographed standing stiffly and staring straight into the camera. She had a better idea. She, the schoolteacher, would hold a book, and she would sit on the stone wall in the shade of their big apple tree and show her mother how to hold and click the shutter of this new wonder, the camera.

With these fanciful ruminations in my mind, the "Rock Wall" Helen York was sitting on suddenly emerged as a possible key to deciphering the past. I needed to know *which* stone wall she was sitting on. While creating a map of the farm years earlier, I had made an inventory of the three kilometers of rock walls. There was only one chance in about two thousand that I could pick the one-meter section of stone wall she was sitting on. Without knowing which *side* of the wall she was sitting on, it would be doubly difficult to find that spot by random chance.

The photo showed nothing in focus by way of background and behind the tree, and so it gave no clue to the photographer's position. I guessed, though, that the picture might have been taken away from the sun and toward the buildings. If so, I might be able to identify the spot where she sat, if I could recognize the rocks. Rocks are routinely removed or fall from stone walls, but the lower rocks could hardly have shifted, and each one has its own individual shape and relation one to another. It was a reality that at that same time concerned me, because I was rebuilding the foundations of the old Adams/York homestead.

With a print of the picture of Helen "Reading on a Rock Wall at Home" in my hand, I stood where I imagined her photographer had stood facing Helen and the old apple tree. I knelt down, pretending to hold an imaginary camera, and noticed where the surfaces of two rocks fit together at an angle of about forty-five degrees. There was an oval opening in the center of that cleft. I then looked down at the century-old photograph. Helen York had, a century earlier, sat next to that cleft!

I set my chain saw as a stand-in for her onto the stone wall in front of the old tree remnants, positioned myself so I was facing where the old homestead once stood, and snapped a picture. Later, when I set the two photographs side by side, I was stunned — the same distinctly shaped and angled arrangement of rocks below Helen's right hand was repeated in my photograph. Then other rocks around them fell into the same matching pattern (except that the topmost layer of the stone wall was now missing). The spot where she had sat was fifty paces (seventy-five meters) from the farmstead barn. From the shadow of a sharp corner of one rock onto the rock below, I could also deduce that the photograph was taken late in the morning.

I had proof that the picture of Helen York was also a photograph of that now-dead old apple tree. I had found a connection between the tree, its past, and my present life on the Hill, my home. The past clicked up against the present with a resounding *whomp* to make the link in the long chain that reached back to the first settlers of this area, to those who made American history. I realized that home is where there are both knowledge of the past and hopes and plans for the future. Home is therefore always something in the making through shared experiences that linger in the imagination.

The capacity to have memories and emotions is not unique to us. We have that ability in common with every bird and probably every mammal. What differs are their specific sensations, to which we are not privy, and their expressions of them that are sometimes difficult to interpret. Birds express emotions by vocalizations. I had seen possible memory and emotion in a blue-headed vireo that nested the last May in a balsam fir tree thirty meters from my cabin and sang extraordinarily the previous November before starting its migration. It likely had found a place that appealed and was fixing it in its memory, and yearning brought it back come spring when one (or it) nested there. The loon that kept visiting the lake a kilometer or so down the road came to know it and its occupants. Its memories of home returned it from out in the Atlantic Ocean when the days were getting longer again, and they would help it return to specifically this place. And just so, when I look across the field to that old apple tree stump that will soon be no longer, I know that the memory of it will be with me still, and attached to it will be the history of this Hill, this homestead. It will stay, as a deep and private thing.

ON HOME GROUND

There is a place of trees . . . gray with lichen.
I have walked there
thinking of old days.

— Ezra Pound, "Provincia Deserta"

ALTHOUGH THE CABIN IS TRANSFORMED INTO "DEER CAMP" for a couple of weeks each November, my nephew Charlie Sewall and I are driven yearlong by a strong urge to return home. This urge for our annual migration may reach a peak in November, but the preparation leading up to it is constant. During the summer, a stroll in the woods with friends and family when Charlie visits is often a poorly disguised scouting trip for the perfect tree to sit in during hunting season. As fall approaches, we begin to exchange weekly e-mails about food and other essentials. When the big day is finally here, it may take priority over other commitments — it's time to go no matter what, and when we get to camp on York Hill our tradition is underscored with a few sips of whiskey reserved since the previous fall in an old hollow stump with a cover, disguised as a table in our cabin, and with a venison steak from last season taken from the freezer as a "starter" to kick off the first day of this new deer season.

The whitetails are there all year long, but you would hardly know it until fall. Suddenly they are on everyone's mind. Walk down Main Street or into the Farmington Diner for your morning coffee — walk anywhere and meet a friend — and often the first question is "Gut your deeah yet?" Almost everyone aches to be able to say, "Yup — a nice eight-pointer!" but most will likely respond with "Not yet." And then you'll hear about where they've seen a recent antler rub, a fresh scrape under a fir bough, or the leaves pawed on a beech ridge.

During deer season in Maine, you leave from wherever you are and go to your "camp." It's usually a tarpaper shack on a back road in the hilly woods some fifteen to twenty-five kilometers away. Some of your neighbors will be there, too, and you'll stay up long into the night playing cards and drinking whiskey and beer, and the main topic of conversation is the deer you remember seeing as a kid, the old deer stories you heard, and the buck you hope to see tomorrow. No matter how long you stay up, you get up before daylight to start the wood fire and to boil fresh coffee. After the briefest of breakfasts you go out the door under what you hope will be a starry sky. You look up to the Milky Way splashed across the heavens directly over-head. You walk in the dark to reach a favorite spot, maybe a thick branch up in a spruce tree, a stump, or a big moss-covered rock next to a spruce from where you can look down a long slope through the hardwoods. In the not-yet-light, and not-still-dark, you listen with senses hyper-alert, and you wonder if the blood pounding in your ears is distant footsteps. You later hear the first chickadees as they awake, and a flock of finches chatter as they fly over. A raven calls in the distance, a jay cries, and a red squirrel chatters and rustles across the frosted leaves. An hour later you see the eastern horizon blaze orange through a lattice of black tree silhouettes.

This is what I remember growing up in the hilly country of central Maine in the 1950s, and deer season hasn't changed much today.

. . .

My childhood neighbors, Phil and Myrtle Potter, were both trout fishermen in the summer and partridge hunters in October. But their real love was deer hunting in November. One of my first hunting trips with Phil, when I was probably fourteen years old, was to his favorite hunting grounds by the Potters' camp along a brook on a dirt road in Carthage, near the village of Weld. Like most other camps, this one was a one-room tarpaper shack with a bunk bed, a crude table, and a stove made from an empty metal drum. Like every other camp, it was a retreat from civilization; in this case civilization was Wilton, a major town of five thousand people. Having a camp was a foothold in the wild, and going there felt like a return to the tree house or den you had as a kid. The camp was patched together not according to any blueprints or state, town, or other regulations; it was all yours.

Phil and Myrtle's camp was jointly owned or at least used with "Huck" Williams, also from Wilton. On this trip, Myrtle drove to camp and dropped Phil and me off on the way, so that we could cross through the rugged terrain from Route 156 and trek partly over and around Mount Bald, to come out at camp, at the end of the day if we were lucky. Mount Bald is bare and glacier scraped at the top and has a broad ring of red spruce forest below the ledges. An unbroken forest of maple, beech, and birch covers the slopes all the way down to brooks at its base.

Phil carried his 30/06 in his right hand, and I the .22 squirrel rifle I had bought through the Sears and Roebuck catalog with fifteen dollars earned doing Phil's barn chores in mine. As we stepped off the road and I looked up the mountain, I saw the dense dark spruce thickets near the top through the now-bare branches of beeches and maples. Those spruces seen through a light fog seemed mysterious and distant. As we walked through the hardwoods near the foot of the mountain, Phil showed me claw marks of bears, almost always three or four parallel grooves, on the old beech trees.

We saw black, now-healed scratches in the gray bark, and I was excited to see also freshly grooved yellow ones. Occasionally Phil pointed out possible day-old hoof prints indented into the leaves, and then we found an antler scrape on a sapling. A little farther I saw pawed ground where a sprinkling of dark brown earth had been flung back over the recently fallen yellow and orange leaves. In the middle of the disturbed patch of ground was a large clear hoof print. My mind was now aflame with deer. The hours went by and we hiked kilometers, but we saw no deer, nor heard any. Except for the occasional scream of a blue jay, all was almost spookily quiet. I knew, though, that the deer had to be somewhere, if I only looked hard enough through the foggy haze. That evening we met Huck at camp, and we warmed ourselves by the wood stove as Myrtle cooked supper. Stories were told, and it was a day well spent. I had gotten the deer fever if not the deer, and those same woods have beckoned ever since.

I was a sophomore in high school when Phil first loaned me his fallback rifle, a 30/30 lever-action Winchester (which I now own and use), handed me five bullets, and let me loose on my own. I went out across our field in the back to the barn and into the woods beyond. I hunted there for maybe an hour and a half every morning before school, and then again after coming home. I got to know those woods pretty well. They were mostly semi-mature hardwoods sprinkled with a few giant hemlocks. The big trees over the years all had been worked over by pileated woodpeckers. Carpenter ants had eaten out their rotting centers, and some of them were by now mostly hollow. In the fall when the goldenrod and asters were in bloom in our fields, I had followed a strong line of bees into our woods and found a bee nest in one of these hemlocks. Remnants of an old rusted sheep fence braided down the middle of our woods, and next to it grew dense patches of juniper bushes thickly populated with "rabbits" (snowshoe hares). In late November they showed up white, unless the snow was early, in which case they were invisible.

I had dreamed of "getting my deer" for a long time, and when one morning I saw a movement and a patch of brown ahead of me behind the junipers and it morphed into one, my heart leaped into my throat. But I somehow managed to fire and get the deer. All deer are remembered, but the first is most memorable. I told my friends about my deer at school that morning, and they came out into the woods with me after classes to bring it out. My mother snapped a picture, as Bruce Richards, Buddy York, and I — all in crewcuts — posed with it in the farmyard. The snapshot shows me hoisting the front of the pole that held the deer and Buddy the rear. I had succeeded at something that had meaning to my friends and was proud and happy. Now that I was charged with success, the woods were so exciting that I wanted to spend an entire year in them living off the land, maybe even to get a bear.

The most compelling reason to be in the woods, aside from getting a pet crow, was to catch something to bring home to eat. It was hunting, fishing, or beelining. Even if these natural tendencies were soon sublimated by the intrusion of sobering and civilizing cultural influences, they continued to hold sway.

I still hunt the whitetail deer, but now mostly with Charlie, who graduated from Bowdoin College and then got a graduate degree from the University of North Carolina and is now a toxicologist for Merck Laboratories in Pennsylvania. Pennsylvania is overrun with deer, but he drives the twelve hours it takes to come hunt with me in the home woods where we've always hunted.

More seasons than not, Charlie and I don't "get our deer," and on those occasions we imagine that there just aren't any around. A fresh tracking snow usually proves us wrong, though. No small part of the rebuttal to the invisible deer notion is that deer can hear better, smell better, and get around better in thick woods than we can.

But hunting is not just about shooting a deer; it's about the ritual of being on the land, which is the excuse for the long time spent with generally no obvious practical reward to show for it. Hunting is when you get up at 4:00 a.m. — time enough to fix your breakfast and get out into the woods while it is still pitch black, when everything is still except your footsteps on the frosted leaves. You make it to your favorite tree, climb up, and are in place on a thick limb an hour before daylight. And then you listen. You hear an owl or two hoot and then, as it begins to get light, the *churr*ing of a red squirrel, and the chickadees. On one morning of your two weeks of listening and freezing you may have heard in the distance what you thought were footfalls of a deer, perhaps. And your heart pounded because you recognized that sound as different from others you've become familiar with — a mouse in dry leaves, a grouse walking, a red squirrel scurrying. Chances are you never saw the deer. The sound faded, or stopped. But you've seen chickadees and maybe kinglets from up close. You've contemplated every moss and lichen near your perch. You've walked kilometers, and slowly, very slowly, you have become familiar with "the place." Whether or not you hold title to it, it becomes home simply because you get to know it.

Deer come and go, but the experiences of them, seen and unseen, leave memories more lasting than the taste of venison. One of these was when I brought Charlie to camp in 1989. He was then about the same age as I was when I got my first doe, and I took him into the same beech ridges where Phil and I had seen the bear claw marks and buck's pawing prints on the ground under an overhanging spruce branch, and the antler rubs on nearby young trees. Charlie and I had been hunting all morning and gradually working our way up Gammon Ridge, the next rise over from Mount Bald, to check for activity under the red oak trees that grow on the top. When we made it to the ridge top near noon, we sat down to munch on some snacks. We were sitting under the oaks on a shoulder of the hill, at the edge of a boulder-strewn slope. We could see the White Mountains of New Hampshire in the distance, but rock obstructed the view directly below us. The leaves were dry and rustled easily, and after we had been sitting quietly for a while, we started hearing possible footsteps in the distance below. Their rustle in the dry leaves was coming closer. We tensed with excitement and hunched down, frozen into position, and waited. Suddenly, a buck stepped out from behind a boulder, perhaps only nine meters in front of us. Two shots rang out as one from our already-lifted rifles, and we raced down in excitement to find our buck. We gutted it and spent the rest of the day dragging the eighty-kilogram deer out of the woods. Charles Senior, Charlie's father, had the head mounted as a memento. It is now at our camp and reminds us of that day.

Years later. I had hunted until the very last day of the season and still had not seen a deer, when suddenly a big buck jumped up in front of me, and in my surprise I did the unpardonable — I was careless and the one shot I managed to make only wounded him. There had been heavy rains in the past several days before a snowfall, and the brooks had become raging torrents. I had no problem staying on the buck's track because I saw blood on the snow. But then he headed directly for and then into Alder Brook. I took off my boots and all my clothes, held them and my rifle aloft, and waded through the swirling icy water. Chilled through and through when I got to the other side, I dressed quickly and again picked up the spoor. I spooked the buck again, and he headed back to the same brook. Again I undressed and crossed behind him. Only then was his misery ended in death, although my relief in delivering him did not cancel my disgust at myself for wounding him.

Every year is different. The day after Thanksgiving in 2007, the flu hit me the day we arrived at camp. That night the sky was lit up brilliantly, and I saw a shooting star. I saw another one as soon as I stepped outside under the moonless sky at 4:30 the next morning. Then it clouded over, and we hoped for tracking snow. But at 4:00 a.m. the next day, when our alarm clock again jangled and we jumped out of bed, we heard rain — pounding, gushing, and pummeling rain. Still, Charlie and I were out in the woods and in place on our favorite stands an hour before dawn. His is a flat moss-covered rock under a red spruce tree overlooking a long rocky slope, mine a fir tree with thick branches in the notch between two hills.

The torrential rains only occasionally let up to a drizzle during the next two days, and our normally tranquil brook soon swelled and became a torrent. As we perched on our stands we heard, for hours, the heavy pounding on the soggy leaves below. We strained our ears but didn't hear a footstep or a twig snap — but why should we? We couldn't even hear ourselves walk when we climbed down to wander through the dripping woods. We couldn't see far through the gloom and thick sheets of rain, either. When we made it back to camp at the end of each day it was dark, and we took our shots of whiskey neat and chased them down with beer. We lit a propane lamp and got a fire blazing in the camp stove to fry up potatoes with sliced kielbasa in a big black iron skillet and hung our clothes to drip and dry next to the stove.

It got colder after the clouds blew off a couple of days later. Our tree stand shook and swayed crazily in the gusts of wind that roared and whipped all around. We became tree huggers, and cold ones at that, until we climbed down to wander around in our heavy waterlogged boots. Deer sign was scarce; we saw an antler rub on a speckled alder stem that showed yellow and orange — fresh! The sign recharged us with adrenaline, but by the end of the week the three cups of strong coffee and two ibuprofens with buttered toast and jam for breakfast were no longer enough to keep me going.

I wimped out. I told Charlie, "You know, we don't *have* to get a deer — let's save 'em for next year." He agreed. But as I walked down the hill to get in my pickup truck and drive back to Vermont, he picked up his rifle and went into the woods. He stayed for almost another week. Still, the 2007 hunt left Charlie, not just me, without venison. A month later Charlie e-mailed me that he had already bought his 2008 license; he was already counting the weeks until the next year.

A year later, on the 8th of November, Charlie again drove all night from Pennsylvania. He got to camp just minutes before I did from Vermont. We could have spared ourselves a total of thirty hours of driving and gotten deer more or less by our back doors. But that was not the point — this was the Place. Our York Hill. Period.

Charlie unscrewed the cap from a bottle of Scotch, and we had a swig for good luck and good cheer, and then we had just enough daylight left to check for signs. We saw none.

The next day, Sunday, has since colonial times (separation of church and state notwithstanding) been reserved in Maine as a day when "everyone" traditionally goes to church. Since the natural tendency of men is to be enjoying themselves hunting out in the woods sitting maybe on a stump, instead of a pew in church, it was made illegal to hunt on Sundays. Fat chance of us going to church, so Charlie and I planted an oak tree next to the cabin instead.

We were up at 4:30 a.m. on Monday, lit the propane lamp, and got the fire roaring and the water boiling. We had lain awake at night weaving thoughts of our hunting the next day, as spiders weave their intricate webs in the dark. Now we poured hot water through the coffee filter into our cups and sat down on the couch to savor taste and aroma, companionship, and anticipation just before leaving. And like spiders sallying into their webs from their lairs, we were soon ready to walk our favorite trails that are like the spider's drag lines, because they bring us directly and quickly to our favorite spots far out in our territory, to wait for prey. Out we went from the warmth to walk under a dark sky, one behind the other along the one-kilometer path to the three-notch (all bucks, one about 115 kilograms) big fir tree. I'm the better climber and so elected to go up the tree. Charlie went a hundred meters farther to "The Rock" with the two-notch (one young buck, one doe) red spruce next to it. The spruce's roots growing over the green moss-covered rock make a natural seat, with the tree as a backrest. So now we were in the middle of our territory, poised and ready.

It was still too dark for me to read my watch dial when I reached my familiar perch some seven meters up, where I could stand on a thick branch and drape my arms over two others, right and left. There was still another branch to sit on, if I wanted a new position. I was in a gulley with potential views up two hillsides. But the view in the first faint morning light was more for the imagination than the eyes. On the other hand, my hearing was magnified. A barred owl hooted in the distance. The raven pair would soon call as they awakened on their night perch in the pines by the cabin.

Darkness lifted. The ravens gave their first few *quork*s at the pines on York Hill. They launched themselves to fly side by side into the distance down the valley toward Webb Lake, maybe to the mountains beyond. The first finch flocks awoke — I heard gold-finches and the bell-like jingle of a flock of crossbills. The eastern horizon lightened to orange and yellow and I could start to make out the still-green ferns below. Trunks of blown-over fir trees littered the ground. I could see now into the hardwood forest in front of me, and into the still-dark softwoods to my right. I strained to hear faint stirrings, and I could feel the rhythm of my heart and hear the faint squeak of a shrew in the leaf mold below. Two hours later, I heard a short whistle. Charlie had left his rock and was coming toward me down the slope to pick me up so we could head back to camp for breakfast with another coffee. Afterward we made a tour around the hill, and by two in the afternoon we returned to our stations.

It was dark three hours later when we got back to the cabin from our third, the evening, hunt. We partook first of all of the customary liquid refreshments and then made a stir-fry of kielbasa, with potatoes and carrots. Kerry, a friend from Rockland, had just joined us for the next day's hunt, and in anticipation of it our conversation invariably led to hunters' tales. Kerry, who had time enough in the afternoon after arriving at camp to take a walk through the woods, found an antler rub on an alder, and also a paw mark on the ground. As was the norm, he reported the age and size of the buck, where it was likely hanging out, where it would move, and when. His "knowing" about the deer seemed to end only when it came to the exact number of its antler tines. Kerry's admission of imperfect knowledge was perhaps proportional to his sense of the growing skepticism of his listeners. Still, even if we seldom believe our fellow hunters' every word, we enjoy them. The truth is that deer encounters are unpredictable. The stories in the "sports" magazines make you think there is a magic formula to make them appear; there isn't.

After a sound sleep that night, we were ready again and eager at the proper time: 4:30 a.m. We knew where we were going. I would this time take my stand on the Rock where Charlie often waits, and he would sit by a big spruce tree at a slope to the south. The only question was how Kerry would get to the knoll where he had seen his hot sign. He could either start out with me and then head east the kilometer or so through the woods to his knoll, or he could head down toward the brook and then swing west until he reached it from there. He decided on the latter option, because if he came with us and then headed east from our stations, he might, as he reasoned, "mess up any deer approach from that direction." I was thinking the same, and also something else, and therefore added pointedly: "Maybe at eight you can swing west and meet me at the Rock?"

"Agreed."

It had rained during the night, and our footsteps depressed the wet leaves without making a sound. A deer could walk up on us from any direction, at any time. As soon as it was light, two ravens zoomed over and I heard the ripping sounds of their wing beats, chopping the air. One was aggressively chasing the other, and both were soon out of sight. I heard the high-pitched staccato calls that you always hear in these chases, getting fainter and fainter in the distance. Odd, I thought, because the deer-tagging station in Weld had registered only one deer so far, and that was on the first day of the season nine days ago. The gut pile from that deer would be long gone. What was the territorial bird, which was likely trying to chase out a vagrant scout, trying to defend? Did it expect future gut piles in this area? If so, it likely sees deer — and maybe us? The raven's behavior seemed like an omen of good luck. I'm not superstitious; it's just that my mind wanders more freely when there is much to hope for. Optimism is adaptive — it keeps us active without requiring any assurance of success.

I was chilled to the bone already at 8:00 a.m., and the thought of hot coffee became ever more inviting. I shivered and did vigorous isometric exercises like those a bee does to generate metabolic heat and keep the blood flowing. Darn — where was Kerry? I wondered. Maybe I'll hold out for another ten minutes, I thought.

A pileated woodpecker started drumming. It had found a dead tree or limb that resonated with a deep sound. Each drumroll of about two seconds was followed by a silence, and then by another drumroll. This sequence continued regularly for about fifteen minutes. Why was this woodpecker making its music in late fall? Admittedly, the pileated has to prepare for nesting early because it takes the pair a month just to prepare a nest hole, and because it is a big bird and bigger birds take longer to grow up than smaller ones. But this early drumming seemed like premature anticipation of spring.

I had been distracted, and more time had passed. But I stayed because I wanted to know how long the woodpecker would hammer or drum, and how long afterward it would remain quiet. Well, Kerry will be here any minute now, I thought, and also wondered what a man walking through the woods would sound like on this day. From how far would I first hear him walking on the wet leaves? But I got impatient and stood up to leave.

There — what was that? A twig snapping? Yes, faint footsteps, another little crackle. Kerry should be in view any second. No sight of orange, though — yet. There — a fleck of movement — something *brown*. Silence. I saw nothing move. Seconds later — I saw it now — a deer walking. Deer had not been plentiful this season due to a harsh winter in which many starved. The state had therefore specified a "bucks-only" hunting policy in this area. I didn't see antlers (in some deer, such as caribou, both males and females have antlers, but in the whitetail deer, only the males are antlered, and first-year males' are very short, thin, and usually unbranched), so I couldn't shoot.

The deer came closer. Yes, small thin antlers — a young spikehorn. The best eating kind. I slowly raised my rifle, saw him over my open sights, and did the one thing that is both hard to do and easy — pulled the trigger. The deer crumpled on the spot. As I walked over to the downed deer, I could not take my eyes off him. He was a beautiful animal, and I had killed him. I felt a momentary sadness. We as a species have been hunting "forever," and the lure of the hunt is irresistible. We are all on precious borrowed time. This is real, the way it is. I took a deep breath, and the sadness of death made way for the joy of life. Here, at this spot, they melded together.

I waited another minute and then jacked a second bullet into the rifle chamber, lifted the barrel, and pulled a shot off into the air. Nobody gets two shots at a deer that are separated from each other by more than ten seconds, so Kerry and Charlie would now know we had a deer. I looked at my watch: 8:35 a.m. We would all three soon exult with instinctual awe and excitement, and celebrate with a sip of Scotch, before the long hard drag to bring the precious deer, our food, to the campsite. In several days, I would skin him and process every ounce of meat to steaks, roasts, hamburger, and venison jerky.

But before my companions arrived, before the anticipated celebration and libation, I bent down and slit the buck's belly. I reached up to his diaphragm to make a cut so I could dislodge and pull out the stomach and intestines and leave them for the birds. As expected, both Kerry and Charlie arrived on the run, and we dragged the deer to camp.

I returned into the woods that afternoon without my rifle. As I wandered back to where I had gutted the buck, I heard a commotion of ravens. Some were flying in the vicinity and others were already down at the gut pile. They scattered when I came near. I stayed listening to the "yells" vagrant juveniles make when they are near food and are harassed by territorial birds trying to chase out the youngsters from their home territory. Thinking of the many adventurous winters I'd spent studying the birds in these woods and learning more about them than I ever thought possible, I expected these young aggressively chased ravens to return at the next dawn with a pack of reinforcements of juveniles and so to overpower the defending pair.

That evening, as Charlie and Kerry and I retold our version of the hunt, we suddenly heard an eruption of yipping, yapping, howling, and the tremulous harmonies of coyotes' singing. I made sure to get back to the gut pile the next morning before daylight, and to my not-great surprise there was not one scrap of entrails left. Presumably the coyotes got them during the night, unless the ravens beat them to it in the evening. It was early, and I expected the ravens to come in at any time and therefore waited. But none returned. I think the ravens had won.

Our early-season hunt was over. Duty called. However, Charlie and I came back to the cabin later that month, on November 23 in the evening, as a sunset over the mountains turned the darkening blue sky in the west to yellow, greenish, orange, and then crimson. The distant mountains were a filmy gray-blue, and from my perch in a spruce I saw the white streak of a meteor through the dark lattice of black tree limbs. The branches next to me were covered with chalky and dark green lichens that varied in form from flat clinging to hanging filaments. A wind was roaring and moaning through the trees, and they were whipped about and crackling and snapping. Two pine grosbeaks, making high-pitched sweet whistles, perched next to me in an ash. The ash didn't have any more seeds hanging from its twigs, but the birds were eating the stems where they had been attached. I had never seen that before.

That evening our camp was cold, and I fired up the stove with the dry sugar maple wood I had chain-sawed during the summer and stacked in the cabin. Charlie sliced up tenderloin from the spike-horn and, with chopped onions and butter and olive oil, dumped it into our sizzling big black iron skillet. I diced some carrots and potatoes from our garden and boiled them only briefly. As Thoreau wrote, "It is a vulgar error to suppose that you have tasted huckle-berries who never plucked them." We served our "huckleberries" of spikehorn tenderloin on the wooden plank table and they had never tasted better.

Next morning I woke Charlie and we got up to start the fire. The stove was slow to start because it was so cold that the iron sucked the heat out of the fire, but the wood was dry and we did get a fire. Revived by heat, coffee, and the light of the Coleman lantern, we put on layers of clothes and, by 6:30 a.m., as the stars faded, we saw a sliver of a moon up above the reddening eastern sky silhouetted by the still-black pines and twiggy maples. Charlie picked up his rifle and left, but this time I stayed, with pencil in hand, to try to capture some of my night thoughts before they vanished like frightened deer.

After Charlie returned and we'd had breakfast, we wandered to Cherry Hill, where in the 1950s a long-since-fallen barn and crumbling farmhouse had stood next to ancient dying sugar maple trees in a clearing. The clearing had grown in and the farmhouse had vanished, leaving only a cellar hole. But now, from a recent clearcut of the area, we could again see the surrounding hills, the mountains, and the lake.

The more we become familiar with the four square kilometers of diverse forest we hunt around York Hill, the stronger our bond to it, and the better our chances of hunting success. For more than thirty years we've hunted in these ridges, and we know precisely where we are and can communicate certain locations that may seem devoid of landmarks to others. I can tell Charlie, "Meet me at the Moose Pasture at 10:00 a.m.," and we'll come to the same spot from different directions at exactly the same time. The Moose Pasture was once a cleared yarding area for a lumber operation next to Alder Brook. It quickly became a favorite spot of the bull moose to engage in their premating rut activity every October. With the passing years, this wide-open moose pasture became an overgrown thicket that would be nondescript to most other hunters. But we know exactly where it is, how to find it, and why the trunks of the large moose maples are scarred and cherries and birches broken over at two meters off the ground. Other specific locations are sometimes identified by certain events. "Did you see the buck tracks on the south side of the hill, near where Scott Dixon slept?" asks Charlie. I know exactly where that spot is. More than twenty years ago, Charlie's college friend from Bowdoin came up to visit. It was his first time here, and Scott arrived well after dark on a moonless night with no flashlight. He managed to feel his way along the trail but soon strayed off course. After realizing he had no idea where he was and not being able to take two steps without hitting a tree, he made the smart decision to lie down for the night right where he was. Finding a relatively soft area with some grass, he rested his head on the ground and heard what he thought was the hum of a distant train getting closer and closer. However, after a few minutes he realized that he was directly on top of someone else's home — a large underground yellow jacket nest. He then made the even smarter decision to relocate rapidly. At sunrise when he awoke, he found the trail less than a few paces away, made his way up to the cabin, and joined us for a morning beverage.

Our forest is dotted with such locations linked to memories — the "mud trap," the "underground gurgling stream," the "lucky stump," the "feeder stream," the "swimming hole," the "three-notch fir," "pine hill," the "burned land," the "beech ridge," "tall pines," the "hemlocks," the "north rock," just to name a few. Our familiarity and connection to these spots enable us to find our way to and from the cabin in thick fog, blizzards, and darkness. We've learned every footstep on our favorite paths and can anticipate the feel of the ground under each step, and we can recognize certain trees in the dark that will lead us back to the warmth of the cabin, the central "nest" in our true home territory, where familiarity breeds bonding and sometimes even success.

That night, after a day without deer sign, we heard the wind roaring again. It was much louder than before, reminding us of jet engines revving up. When we awoke in the morning it was to an expanse of white! Wet snow was blowing almost horizontally, plastering onto the southwestern sides of trees and stinging our eyes. The fir branches bent from their heavy loads. The wind picked up, and the trees whipped back and forth, shaking off loads of wet soggy snow. Then it started to pour rain. We were heading out on the north path to the three-notch fir and became soaked from the sweat of our exertion from hiking in waterlogged clothes and boots. It was truly bad weather, the proof being that we went back home even though we had enough light to stay out fifteen minutes longer.

There is nothing cozier than sidling up to a warm wood stove, in a dry change of clothes, while the wind is howling outside and hurling buckets of water onto the roof and against the sides of a cabin every second, after having plowed through kilometers of deep snow in icy rain. But the discomforts of the day were bearable, because we always knew where home was, how to get there by the shortest route, and that sitting by the fire and retelling the old deer stories awaited us.

We didn't sleep well on this 2008 Thanksgiving night after our soaking. I had never before heard such heavy pounding of rain on the roof, or such fierce wind. We heard the brook roaring a kilometer down the hill. There must have been a monumental heat input somewhere on the Earth to evaporate so much water into the air and then provide the energy to bring it here. I wondered how it would be possible for any warm-blooded animal to stay alive during this night of freezing rain, with the snow compacted and holding water like a sponge.

We got up in the dark as always. The wind had died down by then. Charlie pulled a soggy dollar bill out of his wallet and pinned it onto a nail in a ceiling log above the stove: "To tag my deer with." (Each person with a hunting license can "tag," or register, one deer each season at the nearest tagging station, which for us is at Jerry's general store in nearby Weld, and the fee is a dollar, and always has been.) And then we were out before daylight for yet another day.

I soon saw several hale and hearty black-capped chickadees. They were flitting along close to the ground under some fir trees and making their soft high-pitched *tseet* calls. I again heard the pileated woodpecker drum from the same area — and possibly from the same tree/log where it had been busy two weeks earlier. Another made maniacal laughlike calls from about a kilometer distant to the southwest. A pair of red-breasted nuthatches hopped sideways and up and down a pine trunk, and one of them was voicing the nuthatches' usual drawn-out nasal twangs. Two blue jays flashed through the fir trees ahead of me, staying silent, but I heard the sweet whistles of pine grosbeaks, the tinkly chittering of flying crossbills, and here and there the *churr* of a red squirrel. I was walking the kilometer of woods on the west side of the Hill, past the big pines, heading toward Charlie sitting under the spruce at the Rock, and in that distance I also crossed six porcupine tracks, ten fresh coyote tracks, and one deer track. It looked as if both the birds and the mammals had survived quite well. Each has its own unique home here where it had found shelter in the night.

Then I found a new track, made by someone not at home here.

The boot track had a different tread from either of ours, and it belonged to a hunter we did not know. It made me feel uneasy. Our net of private paths and knowledge covers our hill and its watershed like an orb web spider's net covers its hunting territory. We are uneasy of others too close by — we avoid them, or they us.

We have notions of where the deer might be or travel, and we have drifted into hunting as a team. One of us "covers" one area while another covers the next. Our cooperative hunting can work only because we are on home ground we know intimately, and because we have it to ourselves. If a stranger walks between us, he might spoil our game. True, we can stay alive without getting our deer. But what might our feelings be if we were, like the spider, or our ancestors thousands of years ago, dependent on getting prey for survival and reproduction?

I returned to where Charlie was waiting on station at the Rock, so we could walk back together to the cabin for breakfast. We had walked south on the north trail only a short distance when, thanks to the remaining snow on the ground, we found the tracks of a deer that had crossed over the tracks Charlie had made earlier. We hesitated a moment, studying the tracks: "Big enough to have antlers," I ventured.

"Want to swing east by the swimming hole and I'll wait by the south big spruce?" Charlie asked.

"OK — it's a plan." All our directions are based on home base, on the cabin. That's understood.

I backtracked a few hundred meters and then started east down the slope. A half-hour later I turned south, meandering up toward Charlie by the large spruce where we had agreed to meet. I saw some coyote tracks, but also fresh deer tracks. With still half the distance to go, I heard a shot where Charlie was supposed to be waiting. One minute later I heard another shot from the same direction. So now I rushed on and found him standing with a smile beside a fork-horn buck.

We had mutually logged about one hundred hours in the woods this November, during which time we saw deer for a total of about ten seconds, but we had feasted our eyes and ears continually and experienced 359,990 seconds of bonding time to the Hill. Bit by bit, almost every piece of it — every nook and cranny, individual trees, ledges, slopes, thickets, and creeks — is becoming attached to memories that make this home. This was one of them.

As we were admiring our spikehorn, we heard a raven, and then it flew over and called out some more. We left it the entrails and then dragged the deer out to have Jerry tag it at the Weld store, bringing along the now-less-soggy dollar bill that Charlie had stuck up onto the ceiling log by the big black stove to dry that morning. This stove, a Star Kineo, may be over a hundred years old, and it is the heart and hearth of our home. It is what makes the whole hunting enterprise possible in the first place. It is the hub, like the spider's web from where she hunts.

FIRE, HEARTH, AND HOME

THE FIRST THING CHARLIE AND I DO EVERY NIGHT WHEN we get back to the cabin is make or revive our fire in the stove. This is also the second thing I do every morning as soon as I get up there now, in spring, summer, fall, and winter. Our big iron stove is centrally located. It is next to a thick white pine tree trunk from the edge of the clearing that is inserted into the center of the cabin. I lift off one of the four round stove covers, put in a strip of birch bark and some dry kindling, light the bark with a match, and a minute or two after the cover is back on, the fire comes to life.

Some truths come to you when you squint and try to peer far into the fog-shrouded future. Others stand up clear and strong in front of you. The role of fire in making a home is one of the latter.

Flash back to fifty years earlier: Bodo Muche, a newcomer to Africa like myself, and I had met a week earlier in Arusha, a small town at the foot of Mount Meru in Tanganyika (now Tanzania). He hired an old tobacco-chewing *m'zee* ("grizzled old man" in Swahili) and his two rugged spear-toting sons, Karino and Mirisho, and together we started at dawn and hiked up on buffalo trails through dense heath and old forest. We reached the mountain's ancient volcanic crater by late afternoon. There was no sign of anyone ever having been there, but we saw plenty of fresh sign of Cape buffalo, rhinos, and elephants. Descending into the crater, we were enveloped in fog and saw low but thickly gnarled trees with huge spreading limbs that were shrouded with dark green cushions of moss and long flowing lichens. Parts of the level crater floor looked like a flower-sprinkled grazing meadow, and we found a copse of woods there next to a shallow pond. We needed water, so we camped there.

We picked our spot under a big tree, and Karino and Mirisho left with their machetes to gather dry wood. A fire seemed an obvious priority, and we wanted logs that would burn a long time. Within the last month (when I was there in 1962), a friend had been gored and trampled by a buffalo, another chased up a tree by an elephant, and I (or at least my scent and sound) had been charged by a rhino that had apparently been nearsighted but frightening nevertheless. Our pond had trampled mud around it from all of these animals coming to drink there.

Everything seemed suddenly less threatening after we got a fire going. We gathered around it and drew closer. I doubt if we would have slept that night if it were not for that fire. We heard occasional splashes and sometimes the breaking of twigs. But we felt safe by the fire. We just assumed, as though by instinct, that it provided safety.

We had not brought enough food with us and had to shoot a bushbuck.

In the evening after we got it, we sat most of one night around the fire to cut it up and to share it by roasting the meat on sticks. During that time we swapped tales. Fire had made this magically wonderful but scary place hospitable, bringing us together, giving safety and food. The setting and the occasion seemed primal. Where else but around the warmth and safety of the flame could we, a million or more years ago, have plotted the hunt, recruited helpers, shared information, planned, celebrated, flirted, all while sharing good food and being congratulated if not admired for being a good provider and/or cook?

Evidence of ancient campfires is hard to find, and even more difficult to distinguish from that of wildfires. However, new methods are allowing us to determine the temperatures of the fires from the remains, and in conjunction with their placements (such as in deep caves), solid evidence now shows that we used fire as far back as at least a million years, predating our species to at least as early as *Homo erectus*. The latest evidence came from the Wonderwerk Cave in South Africa, where bone was burned deep in caves at the typically high temperatures of cooking fires. Richard Wrangham, an anthropologist at Harvard University, posits that fire, which allowed the cooking of tubers and meat and facilitated digestion and therefore provided a new source of high-energy food for us, was an evolutionary tipping point that drove larger brain development and thus spawned *Homo sapiens* from our apelike predecessors.

When a troupe of chimps chases a monkey up a tree where it has only a limited number of escape routes, the troupe can surround it to cut off those escape routes and then catch it "by hand." However, such hunting is not possible in open habitat where we evolved, where prolonged high-energy chases were required. Furthermore, even if and after we killed an antelope, it would have been visible and could have been taken from us by the much stronger carnivores, lions, leopards, hyenas, dog packs. In order for us to utilize the prey, we may have had to bring it "home" to keep and to cook it, in a nesting place where the family resided. It seems likely that no real hunting culture among apelike creatures could have been possible until they had solved the problem of a haven, such as a cliffside, or cave. A fire there would have sealed the deal, because it would have been used to guard the entrance. A fire, though, would have to be tended, and the safe home it provided would have, as in mole rats, bees, termites, and all other social animals, provided the means for division of labor. The strongest runners chased the prey. Others who were less able could now stay home for their own safety, and to provide service by minding the kids, tending the fire, and processing and cooking the meat. The hearth became the home.

Keeping fire, and presumably much later also creating it, would have been a precious skill. No other animal has ever mastered it. But before we invented the tool to make and then catch a spark and turn it into a fire, we had to "understand" fire. We had to know its behavior, its quirks and characteristics. Otherwise we would have had no idea what to do with a spark, to make it grow into a tiny flickering flame, and to tenderly nurture it, keep it alive, and make it grow. These skills are not rote. They are built on understanding or "empathy" for the fire, as though it were alive. Our understanding of fire required something akin to what psychologists and behavioral biologists call "theory of mind," the capacity mentally to place ourselves into the life of another being or thing, which allows us to predict how it might react. A possible precursor of that capacity may have been our intense social nature, which was also a precondition for our hunting tradition requiring us to track animals. The fire then allowed us even more opportunity to be social and to learn from others, which, with the new nutrient boost of meat and fats, promoted brain growth to create an unprecedented spiral of evolution of intelligence.

One of the most important things we had to learn about fire, even before we learned how to create it, was to keep it enclosed and isolated as though it were a dangerous beast. Fire is obedient, but only if we adhere strictly to certain rules. If we do, it makes life under otherwise severe conditions possible. We learned to enclose fire to control it, and for tens of thousands of years its place was central in our homes, usually held between slightly separated rocks. Now we have learned to enclose it in a metal box. Judging from the many guests at my cabin, it seems that few of us are capable of handling fire even when it is enclosed in a stove. Even when caged, a fire still has a strong tendency to punish by producing suffocating plumes of black smoke, or by not consuming the fuel offered and then simply dying.

Our forebears were undoubtedly well aware of the irritating habit of smoke to seemingly follow you when you spend much time near it. To help make the smoke go up you need to build baffles around it; you essentially build a house around the fire and leave a hole at the top for the smoke to escape. In this way the fire is less easily put out by rain, the smoke escapes from the top rather than into the area where you are sitting, and you retain more of its heat. Our hearth has traditionally been three stones arranged in a triangle, which may be placed in the center of a surrounding structure made from brush, animal skins, bark, or sod. We learned that we could burn the ends of two or three logs and keep pushing the ends inward, to feed the fire to keep it alive.

What better illustration of this method is there than the tipi, perhaps the best home that has ever been invented? The Native Americans, in their highly original design, were able to capture an open fire in several buffalo hides on a few lodgepole pine poles. The smoke went straight up and out a smoke hole which, with two projecting earlike flaps, could be closed during rain. The tipi was used by the Arapaho, Arikara, Assiniboine, Blackfoot, Cheyenne, Gros Ventre, Kiowa, Mandan, Pawnee, Plains Cree, as well as the Sioux, and was adopted by the early white explorers in western America, including Buffalo Bill, Jim Bridger, and Kit Carson. The tipi was easy to make, and easy to disassemble and carry to and re-erect at a new place. It was warm inside it in winter, cool in summer, well ventilated, and was sometimes artfully decorated, but its essential feature was that it contained a central fire. Although especially designed for a roving life following the buffalo herds, the tipi was also used for "permanent" settlements and for overwintering on the prairies. It is currently still in style with those who appreciate its beauty and practicality.

For most of our history, homes, if we built any, were like a cage to hold and keep our fire and/or its heat. Eventually with larger homes we had a "fireplace," first by cementing our rocks around it, and then by putting it into a cast-iron cage (the first iron stoves began to be used in quantity around 1728). However, enclosing the fire to some extent defeated its purpose, since the smoke and heat were not easily separated. After Benjamin Franklin in 1741 invented a stove with an enclosed fire where the fumes were circulated around a baffle to yield more heat with less smoke than in an open fireplace, the physical contact between fire and home started to become divorced in parts of the Western world. Recent technological advances in design have greatly helped to capture more heat and to direct the smoke away and out in a flue and then a chimney. And now, with central heating, and outdoor wood burners, the hearth and the home have become significantly separated. However, we may still maintain symbolic semblances with fake fires, and thus the artificial has become real.

All over the world people still gather around "the hearth," so named for the Greek goddess Hestia. We are the animal that has learned to make, and live with, and use, fire for much more than cooking. Previous claims that we are separate from other animals by our ability to reason, to have emotional lives, to make tools, to anticipate, have fallen by the wayside in the path of science. Association with and control of fire, though, still stands as a unique *Homo* trait. And so it may be a, or even *the,* defining human trait, since at least (somewhat arbitrarily) four designated human species, *H. erectus, H. neanderthalensis, H. floresiensis,* and *H. sapiens,* apparently used it.

. . .

A fire, depending on how skilled we are in controlling it, is a centering place to hold us around it. When we could not carry it with us, we would have been bound to it; we could not go far from it. However, after we came to depend on fire, we eventually would have learned to control it to the point that we could carry it with us, such as in coals surrounded by ash enclosed in perhaps wet hide or bark.

The campfire almost literally made our home a movable one. Perhaps it permitted us to leave the womb of Africa and to spread out and eventually inhabit the planet. A fire allows us to go to distant, seemingly inhospitable places because it can be our castle, kitchen, social center, and school.

Jack London in his famous story "To Build a Fire" dramatized how important the control of fire is for human life in the north in winter. I loved his story when I was a teenager, because there was scarcely anything I liked to do better than go out into the woods in the winter, build a fire, and roast something on it, preferably with some of my buddies. The fire allowed us to get and stay away from our domineering housemother at our boarding school. But ultimately, most of the globe (except for a lowland belt around the equator) would have been uninhabitable by us were it not for our use of fire. Perhaps this "mobile home" allowed us to spread around the globe, because it gave us the means to home almost wherever we went. It was not the reason, but the means.

Modern genetic techniques have revealed the entire human population before 1.2 million years ago was a mere eighteen thousand five hundred individuals. We were then restricted to Africa and had remained homebodies for hundreds of thousands of years. Yet finally, only about fifty or sixty thousand years ago, perhaps due to some invention of home-making (such as making fire?), a small band of us left Africa. We know it was only a few individuals, because the geographic genetic patterns of the current human species show a remarkable similarity, despite superficial differences of "race" that developed later. This small population originating in central Africa, after spreading to southern Africa and crossing north to leave Africa, came to and through Europe and Asia, and from Asia one branch spread across the Pacific and into Australia and New Zealand while another crossed the Bering Strait to Alaska during or before the last Ice Age ten thousand years ago. From there these by-now Asians, soon to become Paleo-Indians, spread quickly over the North American continent and also quickly moved all the way down through South America to Tierra del Fuego. In terms of our human history, this settlement has struck many as a mass migration of a restless herd, and it does not seem to jibe with our "other" biological identity as homebodies. But let's think about that.

Almost all over the world, most people live generation after generation in the same locality, because almost nobody wants to go elsewhere when home is as good as it can get. Even in America, perhaps through a filter of restlessness, as most of us are descendants of people from other continents, those who have found a good home tend to stay there. At the boundary of my hometown, Wilton, in Maine, is a sign that reads "Wilton, a good place to live, work and play." And people have always stayed here, if they could. The surnames on the gravestones of the original settlers over two centuries ago are much the same as those of the people who are still here. So if we humans are so prone to stay in place, how can one reconcile our species' takeover of continent after continent, until there was hardly a tiny island in the vastness of the Pacific Ocean that we didn't colonize? If there are huge advantages to staying home, why did we risk dangers to spread widely? It doesn't make sense — or does it?

We don't have to look far into history to see clearly that the spread of our species all over the globe was not just an odd ancient phenomenon. It is one that is still in plain sight. By far the majority of the people living in my town, and in every town in North and South America, Australia, and New Zealand, are derived from recent immigrants from thousands of kilometers away. Millions have streamed to America from Europe, Asia, and Africa, and the migrations are continuing at a rapid pace. Both the Chinese and the Romans erected massive walls to stem the flow, and we are now using the latest technologies to accomplish the same. True, we now have steamboats, railroads, and jet planes and means to get to and on them. But maybe that is the point: we have not changed, but we have quite suddenly invented new *means,* and that has made the spread possible.

We are not dealing with an either/or reality. We are not either wanderers or settlers. The example of the migratory locusts and many other insects may be extreme, but it does show us that in one situation animals are homebodies, and in another they are not. All depends on costs versus benefits, and means. In insects, the same individuals that we would in no way classify as migratory may be suited to wander at one age of their development and be strictly settled down in one place in another; the juvenile stays to feed and grow while the adult has wings and can and does wander far. The same separation of two behavioral tendencies simply by age applies to us. In general, young adults tend to have the travel itch, and they explore. There are advantages to some to move on, and for others to stay. We are like the frogs that reproduce in a puddle: those that strike out may find a lake and start a new population there, or perish. It all depends on what might be out there that we hope or expect to find. But why leave when we have found a home? Is it just by pull, or is push involved as well?

We are like the birds that stay at home all winter when they can, but leave when they have to. We differ from them in one major respect, though. The chickadee has a very specific habitat requirement. Take it out of that habitat to which it is specialized and it dies. Provided we have access to the hearth or its equivalent, we can live

almost anywhere, from the Arctic deep freeze to innumerable isolated tropical islands, and we don't need to come back. But again,
why did we not just stay on our home island rather than leapfrog-
ging into the known dangers of the sea, where untold numbers of
us undoubtedly perished?

We will never know the precise reasons why the first group of
humans left Africa. But there are clues in our genome concerning
those who stayed, and they could suggest why the others left. As
we reinvaded Africa, we would have encroached on our forebears
and likely displaced them in many places. I suspect, though, that
a recent study by Stephen C. Schuster and Webb Miller from the
Pennsylvania State University Center for Comparative Genomics
and Bioinformatics with forty-seven other authors from sixteen different institutions sheds some light on this question. The forty-nine
authors generated the complete genome sequences of five African
group identities, including a Khoisan (!Gubi) and Bantu (Arch-
bishop Desmond Tutu) from southern Africa. The Khoisan (also
known as San or Bushmen), they conclude, are "genetically different from other humans" and, furthermore, "more different from
each other than, for example, a European and an Asian." The Bushmen are apparently near or at the root of the human genetic tree
whose branches extend now to the most recent ones of the natives
in Tierra del Fuego at the southern tip of South America, where the
diversity of human DNA is lowest. The Bushmen's genetic difference is related to their more ancient history. Their mutations were
retained in the population, whereas small bands of humans who
left Africa would have had only a select sampling from the variety
of the genome, so that the resulting splinter population would be
more genetically homogeneous. I believe, however, that it is significant that the Bushmen as a whole did *not* leave their home in Africa
and retained their uniqueness. Why?

Elizabeth Marshall Thomas, in her writings on the life of the

Bushmen, provides clues: "They are a roaming people and therefore seem to be homeless . . . each group of them has a very specific territory which that group alone may use, and they respect their boundaries rigidly . . . the people who live there know every bush in it where a certain kind of food may grow — where there is a patch of tall arrow grass or a bee tree. . . ." In other words, they are precisely the opposite of homeless. Thomas writes further that the Bushmen's "sleeping hollows were like the soft grass nests of pheasants, hidden in the leaves," and that they were "at home" where few would venture. The Bushmen's home is the land they live on, and their campfire is their hearth and home. There was little need to enclose it with a structure; they were free. Living in southern Africa, the Khoisan were thousands of kilometers southward of fierce pastoralist tribes, which could not penetrate their homeland. Being locked in place with no place to go, they had to manage their populations in a land that was marginal to others. As an alternative option to emigrating or escaping, they were so at home that to them everything they needed was at hand, but few others or nobody else wanted their land and so they were not driven out.

Ironically, we are becoming like the vast shiftless herd that becomes threatened by its own adaptations. Our cities "evolved" in the Neolithic for some of the same reasons that large pigeon roosts did, in part for defense against raiders. At the same time, cities became filled with riches, and thus also became targets of predation, in this case from our own kind. Ever-bigger cities meant the mustering of armies and ever-better *defenses* but at the same time engendered improvements in *offenses*. As with trees in a forest in their life-or-death competition to grab the light, human civilization is subject to limits in growth, lest it become vulnerable to collapse from its own weight.

It could be debated whether we "migrated" all over the globe by push or by pull, although I'm not sure the difference matters. It's not a great difference if we're pushed from a place we've degraded or pulled to a better one that is replete with resources. One thing is sure, though: we have almost no innate homing navigation mechanisms. Compared to an albatross, a loggerhead turtle, or a monarch butterfly, we can hardly walk in a straight line except between landmarks. Our lack of innate navigational ability, aside from the use of local landmarks, is almost as good as proof that we evolved as homebodies.

Nomadism is a strategy that reduces internal conflict, but at a price. In India, one of the most overpopulated countries in the world, nomadism is a way of life for many people. There are groups that have adapted to it and become specialists at it. However, despite that learned lifestyle, those practicing nomadism would gladly give it up. The nomads live out of their carts for the simple reason that they have no choice. "I will be the most happy person in the world," one Lohar woman told journalist John Lancaster, "if I get some land and a house."

We may argue that since we are highly social animals, we are not subject to such brutal territoriality as loons. But it doesn't wash. Why did people inhabit "all" of the far-flung islands in the Pacific Ocean? Why and how did people get to Easter Island, over three thousand kilometers from the mainland and sixteen hundred kilometers from the nearest inhabited island? By chance and blown off course no doubt, but why chance it? And how many did not win the lottery? Would the Anasazi have "nested" on high ledges where they used retractable ladders to climb up, and made their tiny chambers on cliff edges, if it were not to escape human marauders? Why would others retreat into the far north, where they had to live in bitter cold and darkness for half the year? It is not in the least remarkable that humans were in Tierra del Fuego just a few thousand years after crossing the Bering Strait. How could they not! They were being dispersed and trying to find a good place to make a home. Like the aforementioned loons in northern lakes, and the Canada geese in my bog where at least a dozen fight viciously every spring for possession of the only safe spot to nest on — a beaver lodge — and where only one pair stays and the others must go far to find any place at all, it's all about the matter of the costs and benefits of leaving versus staying and fighting. But if and when people find a home to stay in, the place grows on them because they have invested in it, and once adapted to it they have their families there, one after another, as long as they can.

Wherever we were, though, we were still at home with our "campfires." Regardless of their new forms, we are at least indirectly still huddled around our fire as though we had never left it, and indeed without it, most of the Northern Hemisphere would now be unpopulated by our species.

HOMING TO THE HERD

*Someone said that Brecht wanted everybody to think alike.
I want everybody to think alike. But Brecht wanted to
do it through Communism, in a way. Russia is doing it
under government. It's happening here all by itself without
being under a strict government; so if it's working without
trying, why can't it work without being Communist?
Everybody looks alike and acts alike, and we're getting
more and more that way. I think everybody should be a
machine. I think everybody should like everybody.*

— Andy Warhol, in an interview
with Gene Swenson for *Art News*

The beginning of morality is to see the world as it is.

— Carl Sagan

THE CHOICE BEHAVIOR OF ANIMALS IN THE ENVIRONMENT
that they live in, their home, is the result of biological evolution that
individuals have experienced over long periods of geological time.
In some so-called highly successful species that inundated and altered their ancient environment, the extreme numbers created a
new set of selective pressures, favoring different physiology and
behaviors. In a huge crowd, the nuances relating to specific indi-

viduals become irrelevant. The crowd itself becomes the dominant feature of the environment, and so it (like money for art) becomes a new stimulus. This stimulus is as real as the roar of a storm, the smell of the ocean, or the sight of a forest of pine trees. Oceanic fishes in the open seas feeding on nearly inexhaustible plankton live in immense schools, and there is little for them to home in on except themselves.

Extreme gregariousness is not just a provenance of some oceanic fishes. Certain insects, birds, and mammals also home primarily to the crowd consisting of each other, and not always with satisfactory results, as the French naturalist J. Henri Fabre famously showed in his observations and experiments with the pine processionary caterpillars, which follow silk trails between their communal home, where they are sheltered at night and in the cold, and their feeding areas out on the tree branches. Fabre determined that they followed each other, normally traveling out in the morning and returning at night. But one group of his caterpillars by accident got onto the top rim of an urn, where they continued to circle for seven days straight, making approximately 335 circuits. None abandoned their evolved behavior pattern where the *others*, and not the biologically relevant environment, had become the reference. This is not to say that others are not biologically relevant — because in the vast majority of cases they are their lifeline; temporary encampments of largely anonymous individuals commonly yield benefits to the individuals in them. Crowds of starlings, grackles, red-winged blackbirds, crows, ravens, various species of finches, swifts, and swallows sometimes sleep in huge communal nocturnal roosts. Not only do the roosts provide a pooling of experience and hence an aggregate knowledge of the environment (such as of suitable chimneys or rock crevices for swifts, and of safe places in cities, away from owls, for blackbirds and crows), but the crowd itself dilutes risk both from predation and also by acting as an information center for finding patchy, widely distributed food. However, in most cases, the gregariousness is a seasonal phenomenon, not a permanent state of being.

The most notable for their extreme gregariousness in North America were the Rocky Mountain grasshopper, *Melanoplus spretus;* the passenger pigeon, *Ectopistes migratorius;* the Eskimo curlew; the Carolina parakeet; and the bison. All prospered to become dominant members of their ecosystem, but recently five of them "suddenly" became extinct, and the sixth nearly disappeared but has been resurrected to a token presence through our conscious effort, thanks to intervention at the national level in the nick of time. Since aggregating can be and often is highly adaptive, the extreme examples of animals homing to each other rather than to a place and of that resulting in the ultimate of maladaptiveness — extinction — are of interest.

First, the enduring mystery of the Rocky Mountain grasshopper. This species formed truly massive sky-darkening swarms of sometimes billions; it is only one species of about ten grasshoppers that aggregate, out of the ten thousand other species of the Acrididae that don't form swarms. Large population size is generally considered to be a solid buffer against extinction. What went wrong? There is debate about what happened to the Rocky Mountain grasshoppers; they were gone before we could study them live, although two other species with similar habits are still extant in the Old World, and from them we can get insights into the swarming habit.

The previously mentioned desert "locust," *Schistocerca migratoria,* another of those ten migratory or aggregating hoppers, is probably the best known and best studied of the group. It is of Old Testament fame for coming in clouds and descending on the land to devour every green thing and then moving on, leaving devastation and famine in the land. A second species, the brown locust,

Locustana pardalina, of southern Africa, has the same pattern: it has an aggregating migratory phase, and a sedentary one of another appearance and behavior. The two phases of the same species were the cause of confusion; it was long thought that the migrating locust was a different species.

The scale of size and movement of the "locust" swarms is hard to imagine, given the few grasshopper species that we know locally, which seldom fly and which, when they do, rise up for a few meters only before settling back down onto the ground in some patch of sunshine. Grasshoppers are a favorite food of birds, and most have amazingly cryptic coloration that helps them blend into the background environment in which they live. Most species don't show the slightest sign of being attracted to another. But the migratory desert "locusts" orient to each other, and that is what helps to produce the massive wandering swarms; and because they wander and sweep up all those in their path, the swarms become ever bigger.

A surprising discovery having to do with the desert locusts' aggregating and wandering responses is that these actions are *facultative.* That is, when populations are sparse, the grasshoppers actively avoid each other. It is only when they are dense that the hoppers become attracted to each other. At low population densities, the hoppers are colored green so that they are cryptic in the vegetation, which helps them to avoid detection by predators. They then also have relatively short wings. On the other hand, at high population densities when they start to eat plants with poisons, they start to become less palatable to predators, and they are bright orange and yellow, which warns potential predators to avoid eating them. That is, their diet, appearance, shape, and behavior converge to produce a different adaptation, and each of these two adaptations is suited to stationary versus vagrant life strategies. The major take-home point here is that the behaviors are responses to a specific environmental cue, namely, *each other,* beyond a certain population density. Many other insects also have specific adaptations that aid them to disperse at some point in their life cycles, but in those cases the responses are to stimuli such as food and signals from the physical environment.

The desert locust swarms were, because of their massive size, for a long time relatively immune to destruction by humans despite massive efforts to try to fight and even eradicate them. The main reason for this failure was that the efforts were applied on the swarms. They are too large to be overwhelmed by any potential predator, which was the selective pressure for their evolution in the first place.

The developmental switch that turns the normally solitary cryptic and sedentary grasshoppers into forms that are no longer cryptic and that swarm is of obvious practical interest. We now know that the transition is activated by tactile stimuli. The solitary-phase grasshoppers have one-third more sensory receptors on their hind femurs than do vagrant grasshoppers, and the aggregation behavior can be induced in solitary locusts if the nerve that innervates their hind legs is artificially stimulated. While it could lead to interesting experimental results to immobilize the legs of grasshoppers so that they could not make the developmental transition to the aggregation phase, it won't yield a practical application. Spraying them in their nurseries so that their populations remain low enough not to trigger aggregation and the wandering phase may be.

The Old World desert locust, a species we have now battled with our whole arsenal of science for a century, still exists. But the last live specimen of the American species, the Rocky Mountain grasshopper, was collected in 1902, long before we understood anything about them and long before we could have mounted any deliberate and scientifically effective attack on them. This grasshopper, the only American species known as a "locust," was perhaps one of the most gregarious and destructive grasshoppers of all. It was the most abundant insect on the Great Plains, occurring from Canada to Texas. As documented in the recent book by Jeffrey A. Lockwood, an entomologist from the University of Wyoming who has spent a lifetime delving into this species' past in North America, swarms contained billions of individuals and in terms of overall biomass approached that of the bison herds. Egg numbers per square kilometer of soil on one Minnesota farm were near four million, and there were nearly as many on one in Utah. When a swarm came in the mid-1800s, it was likened to the approach of a cyclone that blackened the skies. The locusts ate everything in their path, reputedly including the clothing on people, and they were considered one of the prime impediments to human colonization of the Great Plains. Observers recorded seeing the adults of a swarm pickled in the brine of Great Salt Lake "by the hundreds of thousands of tons" as the dead formed great walls along the lakeshore.

In Lockwood's fascinating account of modern-day attempts to unravel the mystery of their extinction, one of the long-standing theories needed to be examined first, namely, that Rocky Mountain locusts are not really extinct at all. Do they exist to the present day, mistaken for some other grasshopper that, like the desert and the brown locusts, have different solitary and migratory phases? Maybe conditions recently have not been right to induce the migratory phase; past outbreaks occurred during droughts. The main posit for a possible present existence of *Melanoplus spretus* was that they were in cryptic form, namely, the still-common solitary *M. sanguinipes*. But how could one know? The answer came from studies of their genitalia, and global warming.

Grasshopper males may jump on anything resembling another grasshopper to try to mate. But what they lack in behavioral discrimination to maintain mating fidelity within the species is made up by mechanics. Grasshoppers, like many other insects, have species-specific genitalia. Male and female genitalia of any one species have a unique lock-and-key mechanism that prevents mating with the wrong species.

Global warming is melting glaciers worldwide, and those in the Rocky Mountains are spilling out tons of dead grasshoppers. A hundred years ago Montana had 150 glaciers, but since 1966 eleven of them have completely melted due to the indirect effects of our own growing numbers. In 2012 only twenty-five of those glaciers remained, and Montana's Glacier National Park is estimated to be glacier-free by 2030.

Piles of grasshoppers have been dropping out of the ice at the foot of melting glaciers in the Beartooth Mountains near Yellowstone Park since the 1930s, and Lockwood and his team have sifted through tons of decaying grasshoppers looking for the hard parts, mainly genitalia and mandibles. The parts that they retrieved matched those of whole identifiable Rocky Mountain grasshoppers of (very rare) museum specimens, and so they concluded that the swarms, some of them dating to about four hundred years ago, had indeed been the migratory species. Comparisons with other present-day grasshoppers found no match. Indeed, a recent study of mitochondrial DNA also disputes the idea that the Rocky Mountain locust still exists as *M. sanguinipes* in disguise. The conclusion: the Rocky Mountain locust is extinct.

But why precisely did this grasshopper go extinct, whereas most of the other grasshopper species have not? This mystery is not yet solved, although Lockwood and DeBrey conclude that the advent of farming and perhaps other changes in land use after the demise of the bison are prime suspects. These are excellent reasons for the lack of *outbreaks* of this grasshopper for over a century. But, after the facts are digested and the known causes of mortality assessed, we still can't truly explain the *extinction*. Unlike the case of the passenger pigeon, which went extinct in part because it was hunted, this species was never hunted. Why didn't it just become rare? It is highly unlikely that there is not a tiny bit of habitat suitable for its existence remaining somewhere. What we know for sure is that what went missing is the locust crowd, and then not a single individual of the species that once turned the skies to cyclonelike blackness remained.

On the time scale we need to consult to arrange our affairs, it was only "yesterday," in 1878, that the last of the immense communal nesting places of the passenger (from the French word *passager*, "to pass by") pigeon were recorded. It was near Petoskey, Michigan, and this nesting area encompassed an immense expanse — over four hundred square kilometers. Before that, observers had described flocks that literally darkened the skies and took days to pass. The flocks encompassed billions of pigeons. Passenger pigeons were described "beyond number and imagination." John James Audubon as well as Alexander Wilson, "The Father of American Ornithology," told of one breeding place near Shelbyville, Kentucky, that in 1806 was several kilometers in breadth and sixty-five kilometers long. Wilson encountered a flock from another nesting place ninety-five kilometers from this one that, from its breadth, flight speed, and duration, he estimated to be 385 kilometers long and one and a half kilometers wide and to contain 2,230,272,000 pigeons. In 1813, as Audubon was leaving his house at Henderson "on the banks of the Ohio" to go to Louisville, Kentucky, he saw a flock that was so thick that "light of noonday was obscured as by an eclipse." The birds passed in undiminished numbers for three days in succession. At the breeding and roosting places, limbs and trees crashed to the ground from the sheer weight of their numbers.

The passenger pigeon swarms moved up and down the continent from at least Nova Scotia to Florida, and they were reputed to nest for most of the year wherever and whenever they found food. They were long-lived, said to live to twenty-four years in captivity, and Audubon estimated that their numbers doubled and quadrupled each year, so the idea of them going extinct was unthinkable. They were the most common bird in America, but sadly, as with the Rocky Mountain locust, no scientific studies of even the most basic sort were made of them that might now help to explain their demise. "Overhunting" is a commonsense assumption, as is deforestation, but I believe these are poor or at least simplistic explanations that do not touch the root of the problem, which concerns their biology.

The best account of the passenger pigeon's probably critical home life was given in the *Chautauquan* by "the last Pottawottomi [*sic*] chief of the Pokaton band" as quoted by Edward Howe Forbush. Chief Pokaton had been camping in mid-May 1850 on the headwaters of the Manistee River in Michigan when he awoke one morning to hear "a gurgling, rumbling sound, as though an army of horses laden with sleigh bells was advancing through the deep forest toward me." Soon he concluded that instead it was the "distant thunder" of an approaching storm, "yet the morning was clear, calm, and beautiful. Nearer and nearer came the strange commingling sounds of sleigh bells, mixed with the rumbling of an approaching storm." Then he beheld moving toward him "in an unbroken front millions of pigeons" that "passed like a cloud through the branches" and also surrounded him and even landed on him. They were mating and preparing to nest. He states that this was an event he had long hoped to witness and so he sat down to watch carefully.

In the course of the day after a great mass of birds passed by him, the trees were filled with birds sitting in pairs that gently fluttered their half-closed wings and uttered the bell-like wooing notes that he had mistaken for the ringing of bells in the distance. But "on the third day after this, the chattering ceased and all were busy carrying sticks with which they were building nests in the same crotches of the limbs they had occupied as pairs the day before." Their nests were finished and eggs laid by the morning of the fourth day, and the hens sat on the nest while the males left to feed and to return by 10:00 a.m., when the females left the nest and the males took over the incubation. In midafternoon, the females returned and again resumed their second bout of incubation while the males left once more to return at sundown. (The closely related and very common mourning dove has the same routine.)

Chief Pokaton found the nesting grounds strewn with eggshells, "convincing [him] that the young had hatched," on the eleventh day after the eggs were laid. He likely refers to the beginning of hatching of the colony — most pigeons incubate at least two weeks. Nevertheless, to me the report of the apparently instant beginning of nesting after settling and the resulting synchrony in breeding of such a huge number of birds is truly remarkable, and I think it may be noteworthy and relevant because it provides contrasts with the mourning dove. Mourning doves are also gregarious; they roost together in small groups, and I see them also in small groups at my bird feeder in the spring at nesting time. But they draw the line in gregariousness when it comes to nesting; they nest independently of one another. Pokaton noted that the pigeon parents fed their young for thirteen days, which is similar to mourning doves. The pigeons then left their young to re-nest, again like mourning doves. These observations show that the pigeons differed from the still-extant mourning dove mainly in their extreme orientation to each other in their nesting in dense colonies.

Almost everything in the passenger pigeons' lives revolved about themselves, and as Chief Pokaton recorded, the vast horde nested synchronously, as one. Could passenger pigeons have nested in small groups, or in isolation? Were the vast numbers as such *the* critical environmental cue for triggering their reproductive behavior? We do not know. However, as I thought about this I recalled a lecture I heard at UCLA by Daniel S. Lehrman just before he died in 1972.

Lehrman was a pioneering behavioral physiologist working in the lab (the only place where he could have made the following discoveries) with the ring dove, *Streptopelia risoria*. What he showed then made a huge impression on me. It was this: adult physiology and behavior in birds, and in this case in the dove, are highly sensitive to sensory stimuli. Lehrman showed that if a caged dove saw a courting male, hormones were released from her brain that started to kick in her reproductive cycle, and in each stage of that cycle (from seeing twigs to build a nest, to feeling eggs on the belly to incubate the eggs, to seeing the young to start producing the crop milk to feed them) it was stimulus perception rather than pure instinct as such that guided development which affected behavior. This is not a novel concept, as it was well known from the pioneering work by Wigglesworth with his bedbugs, and the previously mentioned developmental switch to migratory morphs in locusts is an equally dramatic example. However, its direct experimental proof in a dove underlies the fact that the stimuli as such are arbitrary but are strict with respect to a species and the specifics of its environment. In the passenger pigeons, that critical stimulus was likely, as in the locusts, the crowd.

Pokaton noted that, again as we know for mourning doves, both sexes secreted crop milk or curd with which they fed their young until they were ready to fly. He reported that they stuffed them with mast "until their crops exceeded their bodies in size," and within two days after their stuffing they became "a mass of fat" and the parent birds then drove them from the nest. A difference from the mourning dove, though, was that the passenger pigeon usually laid only one egg.

As in other birds, brood reduction generally means there has been a food reduction. The adult pigeons were reputed never to gather the nuts and acorns in the nest vicinity, leaving that mast for the young to feed on when they came out of the nest. A reasonable scenario therefore is that the fattening up of the squabs just before fledging, until they became "the mass of fat" that Chief Pokaton described, could have been a critical adaptation related to food depletion around the colony; the young birds needed a large buffer of stored energy before they could engage in long-distance travel to feed themselves as a counter-strategy to extreme aggregation. So, if there was a cost to aggregating, the colonies should not have increased to what seemed to be nearly infinite size. Evolution does not generally produce anything "extra" unless it is a byproduct of something else that is useful. Why then did they aggregate so enormously?

The pigeons could have banded together for a variety of advantages, including the sharing of information about locations of rich but patchy food resources. But once aggregated, the communally breeding flocks would have attracted predators. Yet, in the vast numbers that the pigeons achieved, they would have swamped the ability of local predators to make a dent in them. Native Americans from kilometers around probably camped and feasted at the colonies, so long as they were not intruding on the territories of neighboring tribes. It was only after the white man came that the birds were wiped out. But why did one dove become extinct while the other very similar one, which is a popular game bird still hunted all over the American continent, remain as common as ever?

Naturalist John Burroughs's 1877 book *Birds and Poets* provides what I think may be clues. He writes:

Few spectacles please me more than to see these birds sweeping across the sky, and few sounds are more agreeable to my ear than their lively piping and calling in the spring woods. They come in such multitudes, they people the whole air; they cover town-ships, and make the solitary places gay as with a festival. The naked woods are suddenly blue as with fluttering ribbons and scarfs, and vocal as with the voices of children. Their arrival is always unexpected. We know April will bring the robins and May the bobolinks, but we do not know that either they or any other month will bring the passenger pigeons. Sometimes years elapse and scarcely a flock is seen. Then, of a sudden, some March or April they come pouring over the horizon from the south or southwest, and for a few days the land is alive with them.

Burroughs's observations illuminate a critical part of the pigeons' adaptive syndrome. If the birds nested repeatedly at the same place (as many sea birds do), predators would be able to gather and to multiply, and to make inroads on the flock. By *staying on the move,* by arriving quickly and then again leaving quickly, even as large flock size attracts predators, the pigeons stayed a step ahead and the predators could not accumulate. But if so, what did them in anyway?

Burroughs continues: "The whole race seems to be collected in a few swarms or assemblages. Indeed, I have sometimes thought there was only one such in the United States." He went on to note that

scouting and foraging squads are not unusual, and every few years we see larger bodies of them, but rarely indeed do we wit-ness the vast spectacle of the whole vast tribe in motion. Some-times we hear of them in Virginia, or Kentucky and Tennessee; then in Ohio or Pennsylvania; then in New York; then in Canada or Michigan or Missouri. They are followed from point to point, from State to State, by human sharks, who catch them and shoot them for market.

The nineteenth-century ornithologist E. H. Forbush from Massachusetts wrote:

> Schooners were loaded in bulk with them on the Hudson River for the New York market, and later, as cities grew along the shores of the Great Lakes, vessels were loaded with them there, but all this slaughter had no perceptible effect on the numbers of the Pigeons in the West until railroads were built. . . . Every great market from St. Louis to Boston received hundreds or thousands of barrels of Pigeons practically every season. The New York market at times took a hundred barrels a day. . . . Often a single western town near the nesting-grounds shipped millions of pigeons to the market during the nesting season, as shown by the shipping records.

The simple truth is that the pigeons' nesting aggregations were both too large to escape detection and also not temporary enough to evade modern communication and transportation. Forbush writes that every nesting ground became known and was then "besieged by a host of people as soon as it was discovered, many of them professional pigeoners, armed with the most effective engines of slaughter known."

Here, in a nutshell, is the explanation of why the pigeons' superb adaptation hastened their demise. It was not a flaw originally, because for a long time it worked superbly to protect them. But in the end the large scale of their adaptive "perfection" resulted in overwhelming predation. The pigeon had no home boundaries over which to spread itself and continued to orient only to itself, so it could be everywhere, even to the end.

Their social sense was so strong that it drew the new predator, technologically equipped humans, from afar. It made them not only easy targets, but also easily duped. The commercial pigeon harvesters used huge nets in which they caught thousands at a time. They did this by taking a few flock mates, sewing their eyes shut, and attaching them to a perch. These "stool pigeons" then fluttered in place, and the flock came down to attend to them, and so they were caught and slaughtered by the thousands at once.

The passenger pigeons' progenitors, derived from a common ancestor with the mourning dove, had given up their ancestral homes, which the doves still keep. To the pigeons, the only "home" they knew was in the crowd, and now they had become victims of it. Second, and perhaps more important, even with their guns, nets, and pigs, the human predators could have eaten all they wanted individually, with little or no effect on the pigeon population. However, "the market" extended the exploitation beyond any boundaries, practically to "global" proportions limited only by the length of rail lines. And then, when the crowd was no more, the remainder could not recoup, because they needed the crowd as a stimulus in order to nest.

The passenger pigeon's adaptation was like the schooling adaptation in herring encountering the bubble-net-enveloping hunting strategy of a baleen whale. The only real difference may have been that in this instance the whole population of "fish" was concentrated into the same school. Human communication, efficient long-range means of travel by roads and railroad, and also the lack of territorial boundaries of human predators had tipped the scales to make their adaptation their doom. Even though the pigeons did periodically shift their roost locations, they could never give up their habit of staying with the flock and trying to keep up with it, which had always been their defense.

The last passenger pigeon died in the Cincinnati Zoo on September 1, 1914. But today their close cousin (with which they were often confused), the common and very familiar mourning dove, is as common as ever if not more so. It is a favorite of hunters and is

Mourning dove (left), rock dove / common pigeon (right), and the passenger pigeon (center)

one of the most widespread birds of North America. It may be numerically not far from what the passenger pigeon was once, but because it lives dispersed over the entire continent, its numbers don't seem spectacular.

But there is another pigeon, the rock dove, *Columba livia*, that, in contrast to the now-extinct passenger pigeon, thrives specifically *because of* us and with us, living in almost all cities of the world. There are 290 species of pigeons/doves worldwide, but only this one has taken to living primarily in our cities, the world over. It lives underfoot in busy train stations and city streets and nests on window ledges, bridges, underpasses, and in abandoned buildings. It is sometimes hated because it leaves droppings, and admired and loved by others for its homing ability. Why would one pigeon make a pact with humans for at least already fifteen hundred years, and maybe much longer, and recently even experience a population explosion because of us, and the other on contact with humans, in practically a heartbeat, go extinct?

One answer — the main one — again concerns home-making. Almost all members of the Columbidae make flimsy nests and have two pure-white eggs per clutch. Nests of most species are on the ground or in trees. But the rock dove, a native of the Mediterranean region and North Africa, traditionally made its home on cliffs. It "sees" most of our human dwellings as almost perfectly designed, safe home sites. Human-grown grain and even food scraps, too, mean that the rock dove never experiences a total break in the gravy train.

The rock dove was introduced to America in the early 1600s and immediately took to the streets. It now lives over the whole continent. It is a bird at home on almost all city blocks everywhere, on high rises, tenements, and train stations, nesting on pseudo rock ledges such as steel girders, window ledges and cornices, highway underpasses and bridges. Gustav Kramer, one of the premier students of pigeon homing in the 1940s, would have agreed that they nest on predator-safe nest sites — he died on a cliff while trying to reach a pigeon nest. With a stream of food and perfect home sites that are almost predator free, what more could a bird ask for? Not much.

When I think of the pigeon's amazing life, the first image that pops up in my mind is my recent visit in a European train station with thousands of people hurrying along the corridors, multiple shops in a huge cavern with the rumbling of trains and loudspeakers blaring, glaring lights, sky blocked from view by a huge steel-reinforced canopy, escalators running up and down, and trains coming and going. I don't know where it was. I think it was Berlin, but a similar scene of pigeons walking around literally underfoot among the bustling throng of people could have been in Boston, Barcelona, Paris, Moscow, or Rome. These birds are not pets. They are wild birds, as wild as any passenger pigeon ever was. They are social like most pigeons, but they are not dependent on a crowd. They are free. Are we?

Home is where what you do has consequences, and where you expect and get feedback — both positive and negative — from what you do. That feedback is perhaps the main, if not only, mechanism that maintains balance with the environment that we deem relevant to us. Not being homebound permits exporting the costs of our actions. Without feedback, either positive or negative, we inevitably enter what biologist Garrett Hardin famously called "The Tragedy of the Commons." It's the use of the oceans as a dump for plastics, the unrestricted fisheries that depleted the fish stocks. It's the impact on the anonymous, who live outside the home boundary. Instead of "immediate" local death for mistakes, we can anticipate long-term global effects instead. Ultimately, however, our unrestricted use of powerful technology has made the whole global environment our home. It's our *oikos*, the Greek word for "house" or home that the German biologist Ernst Haeckel coined for ecology.

We are, like other animals, adapted to invent boundaries that enclose and specify our home so that we can make, improve, and defend it. Home to us is not the "out there" in the far reaches where what we do has no impact, and vice versa. Only knowledge and imagination can reach there. Feelings can follow but do so reluctantly. Sensory experience is more powerful. Seeing the Earth from the moon would be such an experience, and the few of us who have had that experience — astronauts to the moon — report (www.spacequotations.com) that it had a transforming effect on them, because from space they saw no boundaries, and the whole Earth suddenly seemed like home. We went to study the moon but saw the Earth instead. Ironically, perhaps, it is our technology that took us to the moon and beyond, and our technology has literally had global effects and therefore made the whole Earth our home, whereas before it was divided by boundaries of separate homes. Now, by looking into the rearview mirror, so to speak, we see the reality of what we did, in a physical image of what is our communal home. Yet, at the same time that we are seeing the whole Earth, we are also psychologically severing ourselves from it. We are becoming an urban species.

Our population grows at around seventy-nine million per year, and with most of the growth in cities, with people interconnected by ever more electronics, we are becoming increasingly more oriented to each other. We are attaching to and becoming emotionally linked to social, political, religious, economic, industrial, educational, and other social factions, as opposed to the mountains, the prairies, the forests, the winds and the weather, the rain and the soil and the oceans and the fish and the birds, insects, bison, and butterflies — all are the ties that had bound us to our planet, but always before locally, to *a* home.

We are loosening the emotional ties to home ground and forming ties to the ersatz, not because we don't want to be tied to home, but because there is less opportunity to make a home. When we became agriculturists, our hunter-gatherer lifestyle — which intimately bound us to our homes of Earth and its plants and animals — changed to the huge detriment of the Earth that nurtures us. Now, our ancient hunter-gatherer heritage in the context of immense populations and globalization has turned those instincts inward onto ourselves instead.

Social orientation evolved and evolves because it aids immediate survival and reproduction. It does not see what the future might hold; neither locusts, passenger pigeons, nor bison had inklings of drastic changes ahead of them from the world they evolved in. We changed their world. Now we are changing *ours* — in ways that could never have been conceived. At our current trajectory, the technology we are riding on to see the world from space is also having a huge impact on it. We could end up completing a circle of seeming self-sufficiency, to become a species of crowd shoppers like the hoppers and the pigeons looking en masse for the next bargain. Will we, like a school of fish confronting a whale, continue our until-now proven survival strategies of massing together, or will we see, and change to something new?

I do not here, after talking about all the amazing beauty of the life on this planet, intend to end on a glum note. The point is, we *can* see the magnitude of what's up. We are different in that way from all of "them" that have just bitten the dust, and unique also from the rest that we have considered inferior to us. We are different also in that the destruction of home boundaries that creates the Commons and Garrett Hardin's "tragedy" of it, can also be an opportunity to pull together to face our *common* enemy: massive overpopulation. I think Walt Kelly's comic character Pogo of the Okefenokee Swamp said it best: "We have met the enemy and he is us."

EPILOGUE

We shall not cease from exploration
And the end of all our exploring
Will be to arrive where we started
And know the place for the first time.

— T. S. Eliot, "Little Gidding"

I RETURNED RECENTLY TO THE TINY HAMLET OF HINCKLEY, Maine, where fifty years earlier, after a good deal of exploring, I had grown and graduated from the Good Will Home, School, and Farm. As a line from the song lyrics to "The Green, Green Grass of Home" says: "The old home town looks the same, as I step down from the train." I didn't know what to expect, as I drove on the highway along the old railroad tracks next to Maine's great river, the Kennebec. Once on campus I walked down the lane past Pike Cottage, where I lived during my six years at Good Will, and then continued on up Uncle Ed's road, where I had been hundreds of times. I wanted to see an old sugar maple tree with the huge limbs where we kids had dangled a long rope from a thick limb. The tree, old even then, was still there, and I now only imagined us climbing the rope practicing hand-over-hand pull-ups, and running and swinging out over the bushes, practicing to become Tarzan. I had not gone to

305

the school willingly but now remembered the place with fondness.

Several of us boys had a secret place out in the woods not far from the rope and had started to make a log cabin. It was our special place where, at different times of the year, we tried maple sugaring and other subsistence living, such as cooking a squirrel I'd gotten with my slingshot over a fire in the snow, or in season roasting a squab on a stick. We'd picked it out of a nest at the barns. On occasion, we even got a porcupine. We tried to escape our cottage and the evil housemother whenever we could, and finally two of my buddies and I decided our outings were so much fun that we had to "run away" and live in the woods permanently (into our foreseeable future, which was not far ahead). Alaska seemed unrealistic, so I convinced my chums to accompany me into the mountains near Weld, a few kilometers from where my family had settled after we arrived in America. I had there felt my first taste of "belonging" and had wanted to stay there forever and become a beekeeper or a farmer. Or maybe it was a trapper. I'm not sure now, but I did want to be somewhere else. As Maine writer E. B. White put it (in "The Years of Wonder" about his trip to Alaska): "There is a period near the beginning of every man's life when he has little to cling to except his unmanageable dream, little to support him except good health, and nowhere to go but all over the place." That was us. And so Phillip, Freddie, and I one night, each carrying a light backpack containing enough food for at least a day, or so, headed out into the woods, in the general direction of the mountains near Weld.

My parents had decided to come to America. They had lost not only home, haunts, and other property, but also parents, friends, a language, social standing, familiar and loved fields and forests. For me on my childhood trip to America, home was in the future, not the past. But it was the recent past that made me want to go back toward the mountains near Weld, near where we had landed after a long unsettled period in Germany.

Papa had grown up on his family farm called Borowke, in what was then the province of West Prussia in Germany. At age seventeen he joined the German army to fight in the First World War to protect his home from the Russian invasion; that was his declared motivation. After his two plane crashes in the newly formed Luftwaffe, and the lost war, his homeland became Poland. But his home was still as ever his home, except he had to convince the now-Polish authorities that he and his forebears had always been loyal to Borowke, to allow them to let him stay home. World War Two loomed, then came, and our family was driven out by the Red Army troops. Traveling with millions of other refugees, after a perilous three-month journey by horse-drawn wagon, truck, train, rickety airplane on an almost empty gas tank, and for a while even a German army tank, and then again by horse-drawn wagon, we finally ended up in western Germany near Hamburg. We were lucky; a farmer there offered us shelter in an open cowshed, and from there we found our one-room hut in the middle of a forest called the Hahnheide. And when we did finally come almost penniless to America and settled on a depleted little farm in Wilton, near Weld, Maine, Papa called it New Borowke, and so I planted a row of trees down the driveway, just like the one he said he had loved at Borowke.

Landing in rural Maine, we were surrounded by small scattered farmsteads. Almost immediately hands and hearts reached out, a party was held, and we met the families of Floyd Adams, Frank Currier, Erland Adams, Phillip Potter, Earl Ellrich, Keith Brooks . . . in a short time we knew neighbors for kilometers around. We were showered with the essentials for living, and our social schedule was constant. When I stepped on a beer bottle along the ditch (I went barefoot in the summer), the town doctor, Herbert Zikel, stitched me up. No money asked. He took us to his camp by Webb Lake next to the mountains in Weld. And when the roof needed fixing, which was immediately, Floyd and Phil came over and helped. My parents needed a horse to haul logs they had cut by hand with a crosscut saw, and neighbor Erland Adams offered them Susie (his tan mare). No money was asked. None given. It was just the neighborly thing to do.

Most of the pioneers in America had come similarly, after severing their ancient roots, presumably because they were no longer flourishing where they were. They may also have sought adventure, to see and experience the new, but I suspect that was mostly in retrospect. They had practically a whole continent to settle in. If they didn't like it one place, they could move on to the next. My family loved where we landed and would not have considered going anywhere else. But many people from there had much earlier left to "go west," which was one reason that Maine woods real estate was cheap. I recall my father thinking that Americans were shiftless, that they had no moral commitment to stay home, but rather wandered like a flock of pigeons to wherever there was mast to feed on. He was afraid that I might turn out that way.

It is true that at that time I needed to explore and wander, as the young of many animals do. I didn't know it then, but my life would mirror that of the albatross and the young salmon that leave their homes to roam, and return as adults to their imprinted home or to the close vicinity of it. The memory that binds does not fade.

The Floyd Adams family had taken us onto their farm and into their home. Floyd took me with his sons beelining in the old apple orchard, and coon hunting with his hound in the woods. He drove us in his maroon Pontiac to show us the nearby mountains and Webb Lake and his camp there along a brook by Holt Hill. On our first drive there we saw a (for me) wonder of wonders perched right over us on a branch of an oak tree: a porcupine. The Adams couple with their three young boys also took us blueberry harvesting on the mountains there, Mount Bald, Tumbledown, and Jackson. We returned home carrying wicker baskets full of berries on our backs. Back at the farm, Floyd had a rowboat on Pease Pond at the edge of their property where we went fishing, and the farm where we settled was on the other side of that pond where Phil and Myrtle Potter lived. They had their own tarpaper shack also near Webb Lake in Weld, in Carthage on the south side of Cherry Hill where Phil took me deer hunting. For canoeing we went on Bog Stream at the village of Chesterville. Western Maine was paradise if you were free to roam, but then when my parents had to leave "to make a living," my sister and I were left for six and eight years respectively at the Good Will School, from which after five years I prematurely tried the aforementioned return home by making a beeline back to that area that I had grown to love. And, after walking two days and one night, we (almost) got there.

I angered my father by "running" away from Good Will School where he had placed me. However, in retrospect, I think he should have praised me, because one of his values, almost in a moral sense, was loyalty to one's home. Two years later I ended up, by some near miracle, at the University of Maine, but then after I also graduated with honors, I think my father's fears materialized when I went to California, getting married and earning a PhD degree at UCLA and then a professorship in entomology at UC Berkeley. I stayed for fifteen years. Didn't I have any "roots"? Didn't I possess the most basic of values? Although my intent was not to please him, I still started commuting back home to Maine and the farm almost every summer with Kathryn, my bride, and our baby daughter, Erica. However, as might be expected, living with my family at my parents' place with our dog, along with my two tame but free and pesky ravens that we had brought along, created friction. We were not as welcome as I had expected; parent-offspring conflict started. I was dismissive and almost oblivious of it then, because I was focusing on my fieldwork on bumblebee behavior and pollination.

In 1976, after I had commuted from California every summer to be home to study my bumblebees, Mike Graham, a Maine neighbor and my former University of Maine proctor and friend at "The Cabins" (housing for independent living reserved for low-income students) where I had stayed when I studied there, told me that "about three hundred acres" of land were for sale near Weld. I didn't hesitate to apply for a loan to purchase this land, which was the old farm on Adams Hill (later also known as York Hill). Four years later my new bride, Maggie, and I lived there one summer in an existing old one-room tarpapered hunting shack with a weathered sign over the door reading "Kamp Kaflunk." For company we had a tame great horned owl and two tame American crows. The owl and the crows roamed free in the woods and sometimes accompanied us the kilometer down to the Alder Brook for mutual baths. We later built a log cabin in the nearby overgrowing field, where the old Adams/York homestead had stood before it was taken by fire in the early 1930s. The foundation of unmortared fieldstones remained but was crumbling into the old cellar hole now sprouting trees. We had ongoing projects studying bumblebees, white-faced hornets, butterflies, and ant lions.

Having bought title to the property in 1977, I was ever more eager to stay home, and so I resigned my position at Berkeley and returned to Maine the first chance I had. That was in 1980, when I was offered a position at the University of Vermont. This was close enough for me to continue the fieldwork, and at the same time to start puttering around in the old cellar hole and dreaming about making a home on the Adams/York Hill farm site. To the albatrosses, loons, and us, home often turns out to be near where we grew up. Memory and yearnings bring us back, and knowledge binds us.

. . .

Like beavers' home-making, mine involved making a clearing. I sharpened my ax and went to work clearing brush. Dark shady forest can be depressing to live in continuously, although I understand the benefits of keeping forest precisely like that, for the animals that need it. The animal in my soul, though, likes sunshine and "a view." The clearing I eventually created is an island in the ocean of forest that surrounds it. It is the only place where the fireweed, goldenrod, and meadowsweet bloom, and so it is a prime place for butterflies and various interesting kinds of wasps, flies, bees, and beetles. Come late summer, bumblebees swarm over the flowers to collect nectar and pollen. White-faced hornets built their paper nests in the meadowsweet and raspberry bushes, and for me they became convenient and cooperative subjects for studying the effects of their high motivation to attack related to their body temperature (it is raised by several degrees before and when they fly at you to sting). Admittedly, not everyone would enjoy these sorts of activities, which require time investment and pain tolerance, but most people would appreciate the birds.

The clearing was and is alive with bird species that live nowhere else near there. First and foremost is a woodcock male (not necessarily the same one) who displays each evening and at dawn in the spring. His spectacular sky dance would entrance anyone who has an ounce of blood in his or her veins, and after the clearing greens up and the red and yellow hawkweed starts to bloom, there are the warbler concerts. The open habitat surrounding the cabin is home of the chestnut-sided, yellowthroat, and Nashville warblers. A pair of American goldfinches, American robins, and cedar waxwings usually live along the edges in the summer. Recently, wild turkeys have come also. But I don't wish to take credit for this bounty; had a brook been immediately adjacent, the beavers would have made this open place instead. Furthermore, they and their descendants inheriting it would have continued to maintain this home for count-less species for centuries. As it is, I now have a ceaseless task to keep the forest from taking it over.

After answering the burning questions I had had about bumble-bees, I switched to trying to solve a simple question about ravens. This took off and turned into almost a life work in which many neighbors, friends, and students participated. Meanwhile, I have shared the cabin and York/Adams Hill every winter for the past twenty-four years with a group of students for full-time natural history study during their semester break. The log cabin we built earlier served them well, but I eventually needed something more private, and better insulated from the cold in winter, if not also from the entry of the unwanted guests, mainly cluster flies, ladybird beetles, deer mice, and red squirrels.

As a child I liked to be in enclosed spaces such as under a leafy arbor, especially if there was a view. I would then fill in "holes" of unwanted spaces with leafy twigs. Later I'd see another appealing spot. Birds in a nesting mood may make tentative nest starts at different locations, as though trying them out before finally com-mitting to one. Inadvertently, I think I did the same in choosing the new cabin site. After examining several tentative home sites far and wide, in the end I picked the old cellar hole that was already next to the log cabin.

I first started taking another look at that old cellar hole full of rocks in the fall of 2007, when my nephew Charlie Sewall came to York Hill for our annual deer hunt. For some reason I started up my chain saw and began clearing the trees and brush that had grown in it. The next summer, our friend Kerry King, an experienced stone wall builder, came to survey the collapsing walls whose rocks were now more visible. He started pulling out rocks, which induced me to shovel a place to fit a flat stone here and another one there, and Kerry found just the right one to fit between them. A piece of the cellar wall on the northern edge took shape, and we started digging out along the western edge, and to look for a big stone for the corner. And so it went, stone by stone. After a while the work of heavy lifting and shoving around of large rocks morphed into an exercise of pattern recognition. The torn-down remains of the old walls had created a large jumbled pile of mostly odd-shaped rocks, and we tried to find just the right one to fit a space here, and another to fit a different space there, our version of doing a jigsaw puzzle. We looked for the biggest stones first, for the base. The jumble of rocks became like a treasure trove crammed with colorful esoteric facts. Some of the oddest-shaped rocks seemed like useless discards, but Kerry reminded me, "We'll eventually find their place." We found some pretty odd empty places and it became fun to find just the

Fitting rocks to build the north wall of the cellar hole

right rock to fill them. There were also long flat rocks to look for, those that would become "bridges" over other rocks, connecting them into and strengthening the structure. Even the little rocks were useful — they became the "fillers."

I liked the way the rocks were fitting together as though they belonged to each other and had the vague feeling that, rather than just playing, we were possibly making a house foundation! The work seemed to me like building one's life, where there are a few basics and bridges, and lots of fillers from a jumble of odds and ends that you don't know at the time what to do with and that may seem obscure. It also felt like the story I was writing about homing, where pieces I had never thought of as being related started to have meaning.

The "foundation" was eventually far enough along that I could envision it half finished, whereas before the work had seemed too formidable to take seriously. I now started to visualize a cabin. A month or so later, while I was on an express train heading to Konstanz, Germany, where I would attend a conference to honor von Frisch student Hubert "Jim" Markl's retirement and hear Randolf Menzel speak about his bee tracking, I daydreamed about someday building a cabin-house. I pulled out my notepad and a pen and made a rough sketch of what I envisioned this house on Adams/ York Hill might become. And so it was that I became more interested in what the history of the Hill had been.

In 2011, long after both Floyd Adams and his wife, Leona, were deceased, I learned to my great surprise that the Adamses for whom *this* hill was once named had been the same family who had sheltered us at their farm when we came to America in 1951, and without whom we would not have ended up in Maine. My first playmates from whom I learned to speak English were Floyd and Leona's sons, Jimmy, Billy, and "Butchie." Floyd's father, I learned sixty years later (at a York/Adams family reunion), had driven their

The finished cabin

cattle up from the farm where we lived near Wilton, to graze on this very same Adams Hill in the summer.

In this same year also, while I was working on this book, we learned to our amazement that my nephew Charlie (Charlie H. Sewall) is a lineal descendant of the Sewall brothers who originally explored this region and who found the initials "Th Webb" inscribed in the bark of a tree along the "pond" they named Webb Lake, which is the centerpiece of the Weld area. My sister Marianne's husband, Charles F. Sewall, had found and photographed the gravestone of an "Esq." Dummer Sewall (1761–1846) in the nearby village of Chesterville, the site of the bog which in the 1970s was central to my field studies of bumblebees and resulted in my first book, *Bumblebee Economics*. My in-laws then learned that a Dummer Sewall from Bath, Maine, had been deeded land in Chesterville, as a reward for his surveying work in the area around Webb Lake, so the gravestone they found of their ancestor in nearby Chesterville may be that of the very same person who originally explored this area which I now call home. I wonder if the Sewall brothers had perchance climbed the Hill for a view in a clearing there, and stood near the site of the same apple tree under which Helen York had posed for her picture while sitting on the stone wall over a hundred years ago. I doubt it, but it is possible. Remains of the stump of the old apple tree under which Helen York sat (see the chapter "Of Trees, Rocks, a Bear, and a Home") are still there.

Most recently the old apple tree gave me one more surprise. It was on August 24, 2012, when at 8:00 a.m. I was walking past it down the Hill to Alder Brook. A wispy smokelike plume was rising from the stump and percolating through the green foliage of the fast-growing maple trees for the eventual sugaring grove. As I came closer, I saw that every centimeter of the huge stump was shimmering. Sunlight was glinting off the moving wings of tens and maybe hundreds of thousands of black ants. They were "alates" (the sexuals) that were leaving their home on their nuptial flight. Among them were also almost as many honey-colored (wingless) workers, which would submerge back into the ground of their home and never leave it alive. Each of the (temporarily — they shed them soon after their flight) winged ants lingered briefly as it came from deep within the ground that is honeycombed with the rotting roots of this old apple tree that had been their home, and then it launched upward, beat its weak wings on its hopeful flight for a mate and a home, and was then blown away by the breeze.

ACKNOWLEDGMENTS

MANY PEOPLE HAVE SHARED MY HOME IN MAINE, AND THEIR love for it and the land surrounding York Hill, as well as their companionship and inspiration, has solidified my attachment to the place. In particular they include Charlie Sewall, Lynn Jennings, Glenn Booma, and Kerry Hardy. I thank the late Albert Sawyer and his son A. Kendall Sawyer and Anne Agan for pictures and valuable historic information about York Hill. The chapter titled "Of Trees, Rocks, a Bear, and a Home" is adapted from an article previously published in *Yankee* magazine. Randolf and Mechtild Menzel, and George Happ and Christy Yuncker, gave me homes away from home on research expeditions that brought me up close to the homing behavior of bees and sandhill cranes, respectively. I am especially grateful for the friendship of Timothy Otter and for his sharing his information on ladybird beetles. I have over the years had input from Lincoln Brower on monarch butterflies, Robert "Swifty" Stevenson on tropical migrant butterflies, Randolf Menzel and Thomas Seeley on honeybees, Theunis Piersma and Robert Gill on shorebirds, Kent McFarland on thrushes, Eric Hanson on loons, David Ehrenfeld on sea turtles, David McPheters and Kerry Hardy on eels, Peter Gillette on star movements, Julie Reid and Ruth O'Leary on geese, and Gary Clowers on a mouse. Douglas Morse and Larry Weber introduced me to spiders and identified Charlotte, Kristof Zyskowski and James Prozek provided leadership and companionship during the trip into the wilds of Suriname. Craig Neff and Parmelia Markwood introduced me to the fascinating works of Maria Sibylla Merian. Andors Kiss and David Russell gave valuable information in answer to various questions. Sandy Dijkstra provided early suggestions on an initial draft proposal for this book. Homing in on the writing then led to twists and turns, providing unanticipated insights and constant temptation to stray. For much-needed focus and refinement I owe my sincere gratitude to the patience and guidance of my editor, Deanne Urmy, and copy editor, Barbara Wood. The ultimate responsibility for this book, though, is strictly my own.

FURTHER READING

Introduction

Albatrosses

Akesson, S., and H. Weimerkirsch. Albatross long-distance navigation: Comparing adults and juveniles. *Journal of Navigation* 58 (2005): 365–73.

Bonadonna, F., C. Bajzak, S. Benhamou, K. Igloi, P. Jouventin, H. P. Lipp, and G. Dell'Omo. Orientation in the wandering albatross: Interfering with magnetic perception does not affect orientation performance. *Proc. R. Soc. Lond. B* 273 (2005): 489–95.

Fisher, H. I. Experiments on homing in Laysan albatrosses, *Diomedea immutabilis. Condor* 73 (1971): 389–400.

Loons

Evers, D. C. "Population Ecology of the Common Loon at the Seney National Wildlife Refuge, Michigan: Results From the First Color-Marked Breeding Population." In *The Loon and Its Ecosystem*, edited by L. Morse, S. Stockwell, and M. Pokras, 202–12. Concord, NH: U.S. Fish and Wildlife Service, 1993.

Mager, J. N., C. Walcott, and W. H. Piper. Nest platforms increase aggressive behavior in common loons. *Naturwissenschaften* 95 (2008): 141–47.

McIntyre, J. W. *The Common Loon: Spirit of Northern Lakes*. Minneapolis: University of Minnesota Press, 1988.

Piper, W. H., D. C. Evers, M. W. Meyer, K. B. Tischler, J. D. Kaplan, and R. C. Fleischer. Genetic monogamy in the common loon (*Gavia immer*). *Behavioral Ecology and Sociobiology* 41 (1997): 25–31.

Piper, W. H., J. N. Mager, and C. Walcott. Marking loons, making progress. *American Scientist* 99 (2011): 220–27.

Piper, W. H., K. B. Tischler, and M. Klich. Territory acquisition in loons: The importance of take-over. *Animal Behaviour* 59 (2000): 385–94.

Piper, W. H., C. Walcott, J. N. Mager, and F. J. Spilker. Fatal battles in common loons: A preliminary analysis. *Animal Behaviour* 75 (2008): 1109–15.

Cranes Coming Home

Sandhill Cranes

Burke, A. M. Sandhill crane, *Grus canadensis*, nesting in the Yukon wetland complex, Saskatchewan. *Canadian Field Naturalist* 117 (2003): 224–29.

Forsberg, M. *On Ancient Wings: The Sandhill Cranes of North America*. Lincoln, NB: Michael Forsberg Photography, 2004.

Krapu, G., G. C. Iverson, K. J. Reinecke, and C. M. Boise. Fat deposition and usage by Arctic-nesting sandhill cranes during spring. *The Auk* 102 (1985): 362–68.

Miller, R. S., and W.J.D. Stephen. Spatial relationships in flocks of sandhill cranes (*Grus canadensis*). *Ecology* 47 (1966): 323–27.

Russell, N., and K. J. McGowan. Dance of the cranes: Crane symbolism at Çatalhöyük and beyond. *Antiquity* 77, no. 297 (2003): 445–55.

Yuncker, C., and G. M. Happ. http://www.AlaskaSandhillCrane.com/. A website with history and excellent photographs of the cranes described in this chapter.

———. *Sandhill Crane Display Dictionary: What Cranes Say with Their Body Language*. Dunedin, FL: Waterford Press, 201.

Human-Guided Migration of Cranes

http://www.operationmigration.org.

Beelining

American Frontier Lore of Bees

St. John de Crèvecoeur, H. *Letters From an American Farmer: Describing Certain Provincial Situations, Manners, and Customs, Not Generally Known; and Conveying Some Idea of the Late and Present Interior Circumstances of the British Colonies of North America*. 1782.

Homing Orientation by Honeybees

Chittka, L., and K. Geiger. Honeybee long-distance orientation in a controlled environment. *Ethology* 99 (1995): 117–26.

Hsu, C.-Y., and C.-W. Li. Magnetoreception in honeybees. *Science* 265 (1994): 95–96.

Menzel, R., K. Geiger, L. Chittka, J. Joerges, J. Kunze, and U. Müller. The knowledge base of bee navigation. *J. Exp. Biol.* 199 (1996): 141–46.

Menzel, R., and M. Giurfa. Dimensions of cognition in an insect, the honeybee. *Behavioral and Cognitive Neuroscience Reviews* 5 (2006): 24–40.

Menzel, R., U. Greggers, A. Smith, S. Berger, R. Brandt, S. Brunke, G. Bundrock, S. Hülse, T. Plümpe, F. Schaupp, E. Schüttler, S. Stach, J. Stindt, N. Stollhoff, and S. Watzl. Honey bees navigate according to a map-like spatial memory. *Proc. Nat. Acad. Sci. USA* 102, no. 8 (2005): 3040–45.

Menzel, R., A. Kirbach, W. D. Haass, B. Fischer, J. Fuchs, M. Koblofsky, K. Lehmann, L. Reiter, H. Meyer, H. Nguyen, S. Jones, P. Norton, and U. A. Greggers. A common frame of reference for learned and communicated vectors in honeybee navigation. *Current Biology* 21 (2011): 645–50.

Riley, J. R., U. A. Greggers, A. D. Smith, D. R. Reynolds, and R. Menzel. The flight paths of honeybees recruited by the waggle dance. *Nature* 435 (2005): 205–7.

Desert Ants

Müller, R., and R. Wehner. Path integration in desert ants, *Cataglyphis fortis*. *Proc. Nat. Acad. Sci. USA* 85 (1988): 5287–90.

Seid, M. A., and R. Wehner. Delayed axonal pruning in the ant brain: A study of developmental trajectories. *Developmental Neurobiology* 69 (2009): 350–64.

Stieb, S. M., T. S. Münz, R. Wehner, and W. Rössler. Visual experience and age affect synaptic organization in the mushroom bodies of the desert ant *Cataglyphis fortis*. *Developmental Neurobiology* 70 (2010): 408–23.

Getting to a Good Place

General Reference

Dingle, H. *Migration: The Biology of Life on the Move.* New York: Oxford University Press, 1996.

Monarch Butterfly Migration

Brower, L. P. Understanding and misunderstanding the migration of the monarch butterfly (Nymphalidae) in North America: 1857–1995. *Journal of the Lepidopterists' Society* 49 (1995): 304–85.

Brower, L. P., L. S. Fink, and P. Walford. Fueling the fall migration of the monarch butterfly. *Integrative and Comparative Biology* 46 (2006): 1123–42.

Etheredge, J. A., S. M. Perez, O. R. Taylor, and R. Jander. Monarch butterflies (*Danaus plexippus* L.) use a magnetic compass for navigation. *Proc. Nat. Acad. Sci. USA* 96, no. 24 (1999): 13845–46.

Froy, O., A. L. Gotter, A. L. Casselman, and S. M. Reppert. Illuminating the circadian clock of monarch butterfly migration. *Science* 300 (2003): 1303–5.

Mouritsen, H., and B. J. Frost. Virtual migration in tethered flying monarch butterflies reveals their orientation mechanisms. *Proc. Nat. Acad. Sci. USA* 99, no. 15 (2002): 10162–66.

Reppert, S. M., H. S. Zhu, and R. H. White. Polarized light helps monarch butterflies navigate. *Current Biology* 14 (2004): 155–58.

Riley, C. V. A swarm of butterflies. *American Entomologist*, September 1868, 28–29.

Urquhart, F. A. *The Monarch Butterfly: International Traveler.* Ellison Bay, WI: Wm. Caxton Ltd., 1987.

Zhu, H., A. Casselman, and S. M. Reppert. Chasing migration genes: A brain expressed sequence tag resource for summer and migrating monarch butterflies (*Danaus plexippus*). *PLoS* 3: e1293 (2008).

Zhu, H., I. Sauman, Q. Yuan, A. Casselman, M. Emery-Le, P. Emery, and S. M. Reppert. Cryptochromes define a novel circadian clock mechanism in monarch butterflies that may underlie sun compass navigation. *PLoS Biol* 6: e4 (2008).

Monarch Butterfly Overwintering

Brower, L. P., E. H. Williams, L. S. Fink, R. R. Zubieta-Hernández, and M. I. Ramírez. Monarch butterfly clusters provide microclimatic advantages during the overwintering season in Mexico. *Journal of the Lepidopterists' Society* 62, no. 4 (2008): 177–88.

Brower, L. P., E. H. Williams, D. A. Slayback, L. S. Fink, M. I. Ramírez, R. R. Zubieta Hernandez, M. I. Limon Garcia, P. Gier, J. A. Lear, and T. Van Hook. Oyamel fir forest trunks provide thermal advantages for overwintering monarch butterflies in Mexico. *Insect Conservation and Diversity* 2 (2009): 163–75.

Slayback, D. A., and L. P. Brower. Further aerial surveys confirm the extreme localization of overwintering monarch butterfly colonies in Mexico. *American Entomologist* 53 (2007): 146–49.

Other Lepidoptera

Chapman, J. W., K. S. Lim, and D. R. Reynolds. The significance of midsummer movements of *Autographa gamma:* Implications for a mechanistic understanding of orientation behavior in a migrant moth. *Current Biology* 59 (2013): 360–67.

Chapman, J. W., R. L. Nesbit, L. E. Burgin, D. R. Reynolds, A. D. Smith, D. R. Middleton, and J. K. Hill. Flight orientation behaviors promote optimal migration trajectories in high-flying insects. *Science* 327 (2010): 682–85.

Haber, W. A., and R. D. Stevenson. "Biodiversity, Migration, and Conservation of Butterflies in Northern Costa Rica." In *Biodiversity Conservation in Costa Rica: Learning the Lessons in the Seasonally Dry Forest,* edited by G. Frankie, A. Mata, and S. B. Vinson, 99–114. Berkeley: University of California Press, 2004.

Orsak, L. J. *The Butterflies of Orange County, California.* Berkeley: Center for Pathobiology Misc. Publ. #3, University of California Press, 1977.

Desert Locusts

Stower, W. J. The colour patterns of hoppers of the desert locust (*Schistocerca gregaria* Forskal). *Anti-Locust Bull.* 32 (1959): 1–75.

Sword, G. A., S. J. Simpson, O. Taleb, M. El Hadi, and H. Wilps. Density-dependent aposematism in the desert locust. *Proc. R. Soc. Lond. B* 267 (2000): 63–68.

Dragonfly Migration

May, M. A critical overview of progress in studies of migration of dragonflies (Odonata: Anisoptera), with emphasis on North America. *J. of Insect Conservation* 17 (2013): 1–15.

Wikelski, M., D. Moskowitz, J. S. Adelman, J. Cochran, D. S. Wilcove, and M. May. Simple rules guide dragonfly migration. *Biology Letters* 2 (2006): 325–29.

Eels

Tesch, F. W. *The Eels*. Oxford, UK: Blackwell Science, 2003.

Dung Beetles

Dacke, M., E. Baird, M. Byrne, C. H. Scholtz, and E. J. Warrant. Dung beetles use the Milky Way for orientation. *Current Biology* 24 (January 2013).

By the Sun, Stars, and Magnetic Compass

General Works on Migration

Berthold, P. *Bird Migration: A General Survey*. Oxford, UK: Oxford University Press, 2001.

Berthold, P., E. Gwinner, and E. Sonnenschein. *Avian Migration*. Berlin, Heidelberg, New York: Springer-Verlag, 2003.

Hilton, B., Jr. Bird-banding basics. *Wild Bird* 5, no. 10 (1991): 56–59.

Papi, F., ed. *Animal Homing*. London: Chapman & Hall, 1992.

Weidensaul, S. *Living on the Wind: Across the Hemispheres With Migrating Birds*. New York: North Point Press, 1999.

Human Spatial Orientation

Cornell, E. H., C. D. Heth, and D. M. Alberts. Place recognition and way finding by children and adults. *Memory and Cognition* 22 (1994): 537–42.

Cornell, E. H., A. Sorenson, and T. Mio. Human sense of direction and wayfinding. *Annals of the Association of American Geographers* 93 (2003): 402–28.

Darwin, C. Origin of certain instincts. *Nature,* April 3, 1873, 417–18.

Doeller, C. F., B. Caswell, and N. Burgess. Evidence for grid cells in a human memory network. *Nature* 463 (2010): 657–61.

Hafting, T., M. Fyhn, S. Molden, M. B. Moser, and E. I. Moser. Microstructure of a spatial map in the entorhinal cortex. *Nature* 436 (2005): 801–6.

Manx Shearwaters

Mazzeo, R. Homing of the Manx shearwater. *The Auk* 70 (1953): 200–201.

Bar-tailed Godwit

Gill, R. E., Jr., T. Piersma, G. Hufford, R. Servranckx, and A. Riegen. Crossing the ultimate ecological barrier: Evidence for an 11,000-km-long nonstop flight from Alaska to New Zealand and eastern Australia by bar-tailed godwits. *The Condor* 107 (2005): 1–20.

Gill, R. E., Jr., T. L. Tibbitts, D. C. Douglas, C. M. Handel, D. M. Mulcahy, C. Gottschalck, N. Warnock, B. J. McCaffery, P. F. Battley, and T. Piersma. Extreme endurance flights by landbirds crossing the Pacific Ocean: Ecological corridor rather than barrier? *Proc. R. Soc. Lond. B* 276 (2009): 447–57.

Piersma, T., and R. E. Gill Jr. Guts don't fly: Small digestive organs in obese bar-tailed godwits. *The Auk* 115 (1998): 196–203.

Pigeons

Fleissner, G., E. Holtkamp-Roetzler, M. Hanzlik, M. Winklhofen, G. Fleissner, N. Petersen, and W. Wiltschko. Ultrastructure analysis of a putative magnetoreceptor in the beak of homing pigeons. *J. Comparative Neurology* 458 (2003): 350–60.

Keeton, W. T. Magnets interfere with pigeon homing. *Proc. Nat. Acad. Sci. USA* 68 (1971): 102–6.

———. The mystery of pigeon homing. *Scientific American*, December 1974.

Schmidt-Koenig, K., and C. Walcott. Tracks of pigeons homing with frosted lenses. *Animal Behaviour* 26 (1978): 480–86.

Somershoe, S. G., C.R.D. Brown, and R. T. Poole. Winter site fidelity and over-winter site persistence of passerines in Florida. *Wilson Journal of Ornithology* 121, no. 1 (2009): 119–25.

Walcott, C. Multi-model orientation cues in homing pigeons. *Integrative and Comparative Biology* 45 (2005): 574–81.

Walcott, C., and R. P. Green. Orientation of homing pigeons altered by a change in the direction of an applied magnetic field. *Science* 184 (1974): 180–82.

Winter Homes

Latta, S. C., and J. Faaborg. Demographic and population responses of Cape May warblers wintering in multiple habitats. *Ecology* 83 (2002): 2502–15.

———. Winter site fidelity of prairie warblers in the Dominican Republic. *Condor* 103 (2001): 455–68.

Morton, E. S., J. F. Lynch, K. Young, and P. Mehlhop. Do male hooded warblers exclude females from nonbreeding territories in tropical forest? *The Auk* 104 (1987): 133–35.

Rimmer, C. C., and C. H. Darmstadt. Non-breeding site fidelity in northern shrikes. *J. Field Ecology* 67 (1996): 360–66.

Rimmer, C. C., and K. P. McFarland. Known breeding and wintering sites of Bicknell's thrush. *Wilson Bulletin* 113 (2001): 234–36.

Townsend, J. M., and C. C. Rimmer. Known natal and wintering sites of a Bicknell's thrush. *J. of Field Ornithology* 77 (2006): 452–54.

Townsend, J. M., C. C. Rimmer, K. P. McFarland, and J. E. Goetz. Site-specific variation in food resources, sex ratios, and body condition of an overwintering migrant songbird. *The Auk* 129 (2012): 683–90.

Bird Celestial Orientation

Emlen, S. T. Bird migration: Influence of physiological state upon celestial migration. *Science* 165 (1969): 716–18.

———. Celestial rotation: Its importance in the development of migratory orientation. *Science* 170 (1970): 1198–1201.

———. Migratory orientation in the indigo bunting, *Passerina cyanea*. *The Auk* 84 (1967): 306–42, 463–82.

———. The stellar-orientation system of a migrating bird. *Scientific American,* August 1975.

Emlen, S. T., W. Wiltschko, N. J. Demong, R. Wiltschko, and S. Bergman. Magnetic direction finding: Evidence for its use in migratory indigo buntings. *Science* 193 (1976): 505–8.

Sauer, E.G.F. Celestial navigation by birds. *Scientific American,* August 1958.

Bird Magnetic Orientation

Edmonds, D. T. A sensitive optically detected magnetic compass for animals. *Proc. Biol. Sci.* 263 (1996): 295–98.

Lohmann, K. J., C.M.F. Lohmann, and N. F. Putman. Magnetic maps in animals: Nature's GPS. *J. Exp. Biol.* 210 (2007): 3697–3705.

Mouritsen, H., U. Janssen-Bienhold, M. Liedvogel, G. Feenders, J. Stalleicken, P. Dirks, and R. Weiler. Cryptochromes and neural-activity markers colocalize in the retina of migratory birds during magnetic orientation. *Proc. Nat. Acad. Sci. USA* 101, no. 39 (2004): 14294–99.

Mulheim, R., J. Bäckman, and S. Akesson. Magnetic compass orientation in European robins is dependent on both wavelength and intensity of light. *J. Exp. Biol.* 205 (2002): 3845–56.

Ritz, T., R. Wiltschko, P. J. Hore, C. T. Rodgers, K. Stapput, P. Thalau, C. R. Timmel, and W. Wiltschko. Magnetic compass of birds is based on a molecule with optimal directional sensitivity. *Biophysical Journal* 96, no. 8 (2009): 3451–57.

Stapput, K., O. Güntürkun, K. Peter Hoffmann, R. Wiltschko, and W. Wiltschko. Magnetoreception of directional information in birds requires nondegraded vision. *Current Biology* 20, no. 14 (July 8, 2010): 1259–62.

Wiltschko, W., and R. Wiltschko. Light-dependent magnetoreception in birds: The behaviour of European robins, *Erithacus rubecula,* under monochromatic light of various wavelengths and intensities. *J. Exp. Biol.* 204 (2001): 3295–3302.

——. Magnetic compass of European robins. *Science* 176 (1972): 62–64.

Wu, L.-Q., and J. D. Dickman. Neural correlates of a magnetic sense. *Science* 336 (May 25, 2012): 1054–57.

Bird Sun Compass Orientation

Kramer, G. Experiments on bird orientation and their interpretation. *Ibis* 99 (1957): 196–227.

Matthews, G.V.T. Sun navigation in homing pigeons. *J. Exp. Biol.* 30 (1953): 243.

Perdeck, A. C. Two types of orientation in migrating starlings, *Sturnus vulgaris* L., and chaffinches, *Fringilla coelebs,* as revealed by displacement experiments. *Ardea* 46 (1958): 1–37.

Genetic Control of Migratory Directions

Berthold, P., and A. Helbig. Changing course. *Living Bird,* Summer 1994, 25–29.

——. The genetics of bird migration: Stimulus, timing, and direction. *Ibis* 34 (1992): 35–40.

Gwinner, E., and W. Wiltschko. Endogenously controlled changes in migratory direction of the garden warbler, *Sylvia borin. J. Comparative Physiology A* 125 (2004): 267–73.

Integrating Orienting Signals

Able, K. P., and M. A. Able. The flexible migratory orientation system of the savannah sparrow (*Passerculus sandwichensis*). *J. Exp. Biol.* 199 (1996): 3–8.

——. Ontogeny of migratory orientation in the savannah sparrow, *Passerculus sandwichensis*: Mechanisms at sunset. *Animal Behaviour* 39 (1990): 1189–98.

Benson, R., and P. Semm. Does the avian ophthalmic nerve carry magnetic navigational information? *J. Exp. Biol.* 199 (1996): 1241–44.

Cochran, W. W., H. Mouritsen, and M. Wilkelski. Migrating songbirds recalibrate their magnetic compass daily from twilight cues. *Science* 304 (2004): 405–8.

Wiltschko, W. U., H. Munro, R. Ford, and R. Wiltschko. Magnetic orientation in birds: Non-compass responses under monochromatic light of increased intensity. *Proc. R. Soc. Lond. B* 270 (2003): 2133–40.

Sea Turtles

Carr, A. The navigation of the green turtle. *Scientific American,* May 1965.

Ehrenfeld, D. W. The role of vision in the sea-finding orientation of the green turtle (*Chelonia mydas*). 2. Orientation mechanism and range of spectral sensitivity. *Animal Behaviour* 16 (1968): 281–87.

Lohmann, K. J., S. D. Cain, S. A. Dodge, and C.M.F. Lohmann. Regional magnetic fields as navigational markers for sea turtles. *Science* 294 (2001): 364–66.

Lohmann, K. J., J. T. Hester, and C.M.F. Lohmann. Long-distance navigation in sea turtles. *Ethology, Ecology, and Evolution* 11 (1999): 1–23.

Lohmann, K. J., and C.M.F. Lohmann. Detection of magnetic field intensity by sea turtles. *Nature* 380 (1996): 59–61.

Lohmann, K. J., C.M.F. Lohmann, L. M. Ehrhart, D. A. Bagley, and T. Swing. Animal behavior: Geomagnetic map used in sea turtle navigation. *Nature* 428 (2004): 909–10.

Luschi, P., G. C. Hays, C. D. Seppia, R. Marsh, and F. Papi. The navigation feats of green turtles migrating from Ascension Island investigated by satellite telemetry. *Proc. R. Soc. Lond. B* 265 (1998): 2279–84.

Papi, F., and P. Luschi. Pinpointing "Isla Meta": The case of sea turtles and albatrosses. *J. Exp. Biol.* 199 (1996): 65–71.

Smelling Their Way Home

Scent Orienting by Insects

Fabre, J. H. "The Great Peacock Moth." In *The Insect World of J. Henri Fabre.* Boston: Beacon Press, 1991.

———. *The Life of the Caterpillar.* 1878. First published in *Souvenirs Entomologiques.* English translation by Alexander Teixeira de Mattos. London and New York: Hodder and Stoughton, 1912.

Steck, K., M. Knaden, and W. S. Hansson. Do desert ants smell the scenery in stereo? *Animal Behaviour* 79 (2010): 929–45.

Mice Homing

Hamilton, W. J., Jr. *American Mammals.* New York and London: McGraw-Hill, 1939.

Salmon Scent Orientation in Homing

Dittman, A. H., and T. P. Quinn. Homing in Pacific salmon: Mechanisms and ecological basis. *J. Exp. Biol.* 199 (1996): 83–91.

Hasler, A. D., and J. A. Larsen. The homing salmon. *Scientific American*, August 1955.

Hasler, A. D., and A. T. Scholz. *Olfactory Imprinting and Homing in Salmon*. Heidelberg: Springer-Verlag, 1983.

Hasler, A. D., and W. J. Wisby. Discrimination of stream odors by fishes and its relation to parent stream behavior. *American Naturalist* 85 (1951): 223–38.

Scholz, A. T., R. M. Horrell, J. C. Cooper, and A. D. Hasler. Imprinting to chemical cues: The basis for home selection in salmon. *Science* 192, no. 4245 (1976): 1247–49.

Procellariiform Birds During Foraging

Hutchison, L. V., and B. M. Wenzel. Olfaction guidance in foraging by procellariiforms. *Condor* 82 (1980): 314–19.

Nevitt, G. A. Sensory ecology on the high seas: The odor world of procellariiform birds. *J. Exp. Biol.* 211 (2008): 1706–13.

Nevitt, G. A., M. Losekoot, and H. Weimerkirch. Evidence for olfactory search in wandering albatross, *Diomedea exulans. Proc. Nat. Acad. Sci. USA* 105 (2008): 4576–81.

Verheyden, C., and P. Jouvien. Olfactory behavior of foraging procellariiforms. *The Auk* 111 (1994): 285–91.

Leach's Petrel Homing and Olfaction

Billings, S. M. Homing in Leach's petrel. *The Auk* 85 (1968): 36–43.

Griffin, D. R. Homing experiments with Leach's petrels. *The Auk* 57 (1940): 61–74.

Grubb, T. C. Olfactory guidance of Leach's storm petrel to the breeding island. *Wilson Bulletin* 91 (1979): 141–43.

——. Olfactory navigation to the nesting burrow in Leach's petrel (*Oceanodroma leucorrhoa*). *Animal Behaviour* 22 (1974): 192–202.

Pierson, E. C., C. E. Huntington, and N. T. Wheelright. Homing experiment with Leach's storm-petrels. *The Auk* 106 (January 1989): 148–50.

Human Imprinting on Scent

Gemeno, C., K. V. Yeargan, and K. F. Haynes. Aggressive chemical mimicry by the bolas spider, *Mastophora hutchinsoni:* Identification and quantification of a major prey's sex pheromone components in the spider's volatile emissions. *Journal of Chemical Ecology* 26 (2000): 1235–43.

Schaal, B., G. Coureaud, S. Daucet, M. Delaunay-El Allam, A. S. Moncomble, D. Montigny, B. Patris, and A. Holley. Mammary olfactory signalisation in females and odor processing in neonates: Ways evolved by rabbits and humans. *Behavioural Brain Research* 200 (2009): 346–58.

Picking the Spot

Communication and Homing by Bees

Beekman, M., R. L. Fathke, and T. D. Seeley. How does an informed minority of scouts guide a honeybee swarm as it flies to its new home? *Animal Behaviour* 71 (2006): 161–71.

Camazine, S., P. K. Vischer, J. Finley, and R. S. Vetter. House-hunting by honey bee swarms: Collective decisions and individual behaviors. *Insectes Sociaux* 46 (1999): 348–60.

Lindauer, M. *Communication Among Social Bees.* Cambridge, MA: Harvard University Press, 1961.

———. Schwarmbienen auf Wohnungssuche. *Zeitschrift für Vergleichende Physiologie* 37 (1955): 263–324.

Seeley, T. D. *Honeybee Ecology: A Study of Adaptation in Social Life.* Princeton, NJ: Princeton University Press, 1985.

———. *Honeybee Democracy.* Princeton, NJ: Princeton University

Press, 2010.

———. Consensus building during nest-site selection in honey bee swarms: The expiration of dissent. *Behavioral Ecology and Sociobiology* 53 (2003): 417–24.

Seeley, T. D., and S. C. Buhrman. Nest-site selection in honey bees: How well do swarms implement the "best-of-*N*" decision rule? *Behavioral Ecology and Sociobiology* 49 (2001): 416–27.

Seeley, T. D., and J. Tautz. Worker piping in honey bee swarms and its role in preparing for liftoff. *J. Comparative Physiology A* 187 (2001): 667–76.

Seeley, T. D., and P. K. Visscher. Quorum sensing during nest-site selection by honeybee swarms. *Behavioral Ecology and Sociobiology* 56 (2004): 594–601.

Seeley, T. D., P. K. Visscher, and K. M. Passino. Group decision making in honey bee swarms. *American Scientist* 94 (2006): 220–29.

von Frisch, K. *Bees: Their Vision, Chemical Senses, and Language.* Ithaca, NY: Cornell University Press, 1950.

———. *The Dancing Bees: An Account of the Life and Senses of the Honey Bee.* London: Methuen, 1954.

Swarm Temperature Regulation

Heinrich, B. Energetics of honeybee swarm thermoregulation. *Science* 212 (1981): 565–66.

———. The mechanisms and energetics of honeybee swarm temperature regulation. *J. Exp. Biol.* 91 (1981): 25–55.

Bird Home Territories

Ahlering, M. A., and J. Faaborg. Avian habitat management meets conspecific attraction: If you have it, will they come? *The Auk* 123 (2006): 301–12.

Amrhein, V., H. P. Kunc, and M. Naguib. Non-territorial nightingales prospect territories during the dawn chorus. *Proc. R. Soc. Lond. B* 271, suppl. 4 (2004): S167–S169.

Bernard, M. J., L. J. Goodrich, W. M. Tzilkowski, and M. C. Brittingham. Site fidelity and lifetime territorial consistency of ovenbirds (*Serus aurocapilla*) in a contiguous forest. *The Auk* 128 (2011): 633–42.

Cornell, K. L., and T. M. Donovan. Scale-dependent mechanisms of habitat selection for a migratory passerine: An experimental approach. *The Auk* 127 (2010): 899–908.

Jones, J. Habitat selection studies in avian ecology: A critical review. *The Auk* 118 (2001): 557–62.

Saunders, P., E. A. Roche, T. W. Arnold, and F. J. Cuthert. Female site familiarity increases fledging success in piping plovers (*Charadrius melodus*). *The Auk* 129 (2012): 329–51.

Bird Nest Sites

Greeney, H. F., and S. M. Wethington. Proximity to active *Accipiter* nests reduces nest predation of black-chinned hummingbirds. *Wilson Journal of Ornithology* 121 (2009): 809–12.

Heinrich, B. *The Nesting Season: Cuckoos, Cuckolds, and the Evolution of Monogamy.* Cambridge, MA: Harvard University Press, 2010.

Architectures of Home

Bird Nests

Borgia, G. Why do bowerbirds build bowers? *American Scientist* 83 (1995): 542–47.

Goodfellow, P. *Avian Architecture: How Birds Design, Engineer and Build.* Princeton, NJ, and Oxford, UK: Princeton University Press, 2011.

Hansell, M. *Bird Nests and Construction Behaviour.* Cambridge, UK:

Cambridge University Press, 2000.

Heinrich, B. *The Nesting Season: Cuckoos, Cuckolds and the Evolution of Monogamy.* Cambridge, MA: Harvard University Press, 2010.

Skutch, A. F. The nest as dormitory. *Ibis* 103 (2008): 50–70.

von Frisch, K. *Animal Architecture.* New York and London: Harcourt Brace Jovanovich, 1974.

Beavers

Aleksiuk, M. Scent-mound communication, territoriality, and population regulation in beaver (*Castor canadensis* Kuhl). *J. Mammalogy* 49 (1968): 759–62.

Barry, S. S. Observations on a Montana beaver canal. *J. Mammalogy* 4 (1923):92–103.

Bradt, G. W. A study of beaver colonies in Michigan. *J. Mammalogy* 19 (1938): 160–62.

Muller-Schwarze, D. *The Beaver: Its Life and Impact.* Ithaca, NY: Cornell University Press, 2011.

Insects

Fraser, H. M. *Beekeeping in Antiquity.* London: University of London Press, 1931.

Hansell, H. M. Case building behavior of the caddis fly larva *Lepidostoma hirtum. J. Zool. London* 167 (1972): 179–92.

Hepburn, H. R. *Honeybees and Wax: An Experimental Natural History.* Berlin, New York: Springer-Verlag, 1986.

Hölldobler, B., and E. O. Wilson. *The Ants.* Cambridge, MA: Harvard University Press, 1990.

Michener, C. D. *The Social Behavior of Bees.* Cambridge, MA: Harvard University Press, 1974.

von Frisch, K. *Animal Architecture.* New York and London: Harcourt Brace Jovanovich, 1974.

Home-making in Suriname

Merian, M. S. *Metamorphosis Insectorum Surinamensium*. 1705. Reproduced in K. Schmidt-Loske, *Insects of Surinam*. Köln: Taschen, 2009.

Todd, K. *Chrysalis: Maria Sibylla Merian and the Secret of Metamorphosis*. Orlando, FL: Harcourt Brace Jovanovich, 2007.

Home Crashers

Myrmecophiles

Pierce, N. E., M. F. Brody, A. Heath, D. J. Mathew, D. B. Rand, and M. A. Travasso. The ecology and evolution of ant association in the Lycaenidae (Lepidoptera). *Annual Rev. Entomology* 47 (2002): 733–71.

Rettenmeyer, C. W., M. E. Rettenmeyer, J. Joseph, and S. M. Berghoff. The largest animal association centered on one species: The army ant *Eciton burchellii* and its more than 300 associates. *Insectes Sociaux* 58 (2011): 281–92.

Wilson, E. O. *The Insect Societies*. Cambridge, MA: Harvard University Press, 1971.

Swallow Bugs

Brown, C. R., and M. B. Brown. *Coloniality in the Cliff Swallow: The Effect of Group Size on Social Behavior*. Chicago and London: The University of Chicago Press, 1996.

Hawk Nest Greens

Heinrich, B. Why does a hawk build with green nesting material? *Northeastern Naturalist* 200 (2013): 209–18.

Charlotte II: A Home Within a Home

Bradley, R. A. *Common Spiders of North America*. Berkeley: University of California Press, 2012.

The Communal Home

Bird Construction Behavior

Collias, N. E., and E. C. Collias. *Nest Building and Bird Behavior*. Princeton, NJ: Princeton University Press, 1984.

Diamond, J. Bower building and decoration by the bowerbird *Amblyornis inornatus*. *Ethology* 74 (1987): 117–204.

Hansell, M. *Bird Nests and Construction Behavior*. Cambridge, UK: Cambridge University Press, 2000.

Heinrich, B. *The Nesting Season: Cuckoos, Cuckolds, and the Invention of Monogamy*. Cambridge, MA: Harvard University Press, 2010.

von Frisch, K. *Animal Architecture*. New York and London: Harcourt Brace Jovanovich, 1974.

Sociable Weavers and Sociable Roosting

Bartholomew, G. A., F. N. White, and T. R. Howell. The significance of the nest of the social weaver *Philetairus socius:* Summer observations. *Ibis* 118 (1976): 402–11.

Walsberg, G. E. Communal roosting in a very small bird: Consequences for the thermal and respiratory gas environments. *Condor* 92 (1990): 795–98.

White, F. N., G. A. Bartholomew, and T. R. Howell. The thermal significance of the nest of the sociable weaver *Philetairus socius:* Winter observations. *Ibis* 117 (1975): 171–79.

Monk Parakeets

Eberhard, J. R. Nest adoption by monk parakeets. *Wilson Bulletin* 108, no. 2 (1996): 374–77.

Hyman, J., and S. Pruett-Jones. Natural history of the monk parakeet in Hyde Park, Chicago. *Wilson Bulletin* 107, no. 3 (1995): 510–17.

Martin, L. F., and E. H. Bucher. Natal dispersal and first breeding age in monk parakeets. *The Auk* 110, no. 4 (1993): 930–33.

Spreyer, M. F., and E. H. Bucher. Monk parakeets (*Myiopsitta monachus*). *Birds of North America* 322 (1998): 1–23.

Van Bael, S., and S. Pruett-Jones. Exponential population growth of monk parakeets in the United States. *Wilson Bulletin* 108, no. 3 (1996): 584–88.

Mole Rats

Alexander, R. D. The evolution of social behavior. *Annual Review of Ecology and Systematics* 5 (1974): 325–83.

Jarvis, J.U.M. Eusociality in a mammal: Cooperative breeding in naked mole-rat colonies. *Science* 212 (1981): 571–73.

Sherman, P. W., J.U.M. Jarvis, and R. D. Alexander. *The Biology of the Naked Mole Rat.* Princeton, NJ: Princeton University Press, 1991.

Social Bees

Evans, H. E. The evolution of social life in wasps. *Proc. 10th Int. Congr. Entomology* (Montreal, 1956) 2 (1958): 449–57.

Evans, H. E., and M. J. West-Eberhard. *The Wasps.* Ann Arbor: University of Michigan Press, 1970.

Michener, C. D. The Evolution of Social Behavior in Bees. *Proc. 10th Int. Congr. Entomology* (Montreal, 1956) 2 (1958): 441–47.

———. *The Social Behavior of Bees.* Cambridge, MA: Harvard University Press, 1974.

Bumblebee Foraging Specialists

Heinrich, B. Foraging specializations of individual bumblebees. *Ecological Monograms* 46 (1976): 105–28.

Oster, G., and B. Heinrich. Why do bumblebees "major"? A mathematical model. *Ecological Monograms* 46 (1976): 128–33.

Desert Mammals

Hamilton, W. J., Jr. *American Mammals.* New York and London: Mc-Graw-Hill, 1939.

The In and Out of Boundaries

Horton, T. Revival of the American chestnut. *American Forests* (Winter 2010). www.americanforests.org/magazine/article/revival-of-the-american-chestnut.

Paillet, E. L. Character and distribution of American chestnut sprouts in southern New England woodlands. *Bulletin of the Torrey Botanical Club* 115 (1988): 32–44.

———. "Chestnut and Wildlife." In *Restoration of American Chestnuts to Forest Lands,* Proceedings of a Conference Workshop, May 4–6, 2004, edited by K. S. Steiner and J. E. Carlson. U.S. Department of the Interior, National Park Service. The North Carolina Arboretum. Natural Resources Report NPS/NCR/CUE/NRR-2006/001.

———. Chestnut: History and ecology of a transformed species. *Biogeography* (2002): 1517–30.

Of Trees, Rocks, a Bear, and a Home

Foster, E. J. *Early Settlers of Weld.* Vol. 1 of *The Maine Historical and Genealogical Recorder.* Portland, ME: S. M. Watson, 1884, 119–123, 172–179.

York, V. *The Sandy River and Its Valley.* Farmington, ME: Knowlton and McCleary, 1976.

Fire, Hearth, and Home

Laubin, R., and G. Laubin. *The Indian Tipi.* New York: Ballantine Books, 1957.

Human Genetics and Evolution

Arjamaa, O., and T. Vuorisalo. Gene-culture coevolution and human diet. *American Scientist* 98 (2010): 140–47.

Huff, C. D., J. Xing, A. R. Rogers, D. Whitherspoon, and L. B. Jorde. Mobile elements reveal small population size in the ancient ancestors of *Homo sapiens. Proc. Nat. Acad. Sci. USA* 107, no. 5 (2010): 2147–52.

Schuster, S. C., W. Miller, et al. Complete Khoisan and Bantu genomes from southern Africa. *Nature* 463 (2010): 943–47.

Stix, G. Traces of a distant past. *Scientific American,* July 2008, 56 –63.

Thomas, E. M. *The Harmless People.* New York: Vintage Books, 1989.

Weaver, T. D., and C. C. Roseman. New developments in the genetic evidence for human origins. *Evolutionary Anthropology* 17 (2008): 69–80.

Homing to the Herd

General

Fisher, L. *The Perfect Swarm: The Science of Complexity in Everyday Life.* New York: Basic Books, 2009.

Leach, W. *Country of Exiles: The Destruction of Place in American Life.* New York: Vintage Books, 1999.

Rocky Mountain Locust

Chapco, W., and G. Litzenberger. A DNA investigation into the mysterious disappearance of the Rocky Mountain grasshopper, mega-pest of the 1880s. *Molecular Phylogenetics and Evolution* 30 (2004): 810–14.

Cohn, T. J. The use of male genitalia in taxonomy and comments on Lockwood's paper on *Melanoplus spretus*. *Journal of Orthopteran Research* 3 (1994): 59–63.

Lockwood, J. A. *Locust: The Devastating Rise and Mysterious Disappearance of the Insect That Shaped the American Frontier.* New York: Basic Books, 2004.

——. Phallic facts, fallacies, and fantasies: Comments on Cohn's 1994 paper on *Melanoplus spretus* (Walsh). *Journal of Orthopteran Research* 5 (1996): 57–60.

——. Taxonomic status of the Rocky Mountain locust: Morphometric comparisons of *Melanoplus spretus* (Walsh) with solitary and migratory *Melanoplus sanguinipes* (F.). *The Canadian Entomologist* 121 (1989): 1103–9.

Lockwood, J. A., and L. D. DeBrey. A solution for the sudden and unexplained extinction of the Rocky Mountain grasshopper (Orthoptera: Acrididae). *Environmental Entomology* 19 (1990): 1194–1205.

Desert Locust

Roessingh, P., S. J. Simpson, and S. James. Effects of sensory stimuli on the behavioral phase state of the desert locust, *Schistocerca gregaria*. *Journal of Insect Physiology* 44 (1993): 883–93.

Rogers, S. M., T. Matheson, E. Despland, T. Dodgson, M. Burrows, and S. J. Simpson. Mechanosensory-induced behavioural gregarization in the desert locust *Schistocerca gregaria*. *J. Exp. Biol.* 206 (2003): 3991–4002.

Stower, W. J. The colour patterns of hoppers of the desert locust (*Schistocerca gregaria* Forskal). *Anti-Locust Bull.* 32 (1959): 1–75.

Sword, G. A., S. J. Simpson, O. Taleb, M. E. Hadi, and H. Wips. Density-dependent aposematism in the desert locust. *Proc. R. Soc. Lond. B* (2000): 63–68.

Pigeons

Bucher, E. H. The causes of extinction of the passenger pigeon. *Current Biology* 9 (1992): 1–33.

Forbush, E. H. The last passenger pigeon. *Bird Lore,* March–April 1913.

———. "Passenger Pigeon." In *Birds of America,* edited by T. Gilbert Pearson. Garden City, NY: Garden City Books, 1917.

Schorger, A. W. *The Passenger Pigeon: Its Natural History and Extinction.* Madison, WI: University of Wisconsin Press, 1955.

Psychology

Weinstein, N., A. K. Przybylski, and R. M. Ryan. Can nature make us more caring? Effects of immersion in nature on intrinsic aspirations and generosity. *Personality and Social Psychology* 35 (2010): 1315–29.

Sensory Stimulation for Nesting in Doves

Lehrman, D. S. Induction of broodiness by participation in courtship and nest-building in the ring dove (*Streptopelia risoria*). *J. Comp. Physiolog. Psychol.* 51 (1958): 32–36.

———. On the origin of the reproductive cycle in doves. *Trans. New York Acad. Sciences* 21 (1959): 682–88.

Lehrman, D. S., N. Brody, and R. P. Wortis. The presence of nesting material as stimulus for the development of incubation behavior and the gonadotropin secretion in the ring dove (*Streptopelia risoria*). *Endocrinology* 68 (1961): 507–16.

INDEX